Project Risk Analysis

With thanks to: Toni Addison, Tara Badder, Tim Bartram, Max Brooker, Phil Brown, Vicky Carter, Louise Chick, Katherine Clarke, Neil Dunkerley, Julia Fedoruchenko, Steve Green, Giles Guthrie, Rob Halstead, Louise Jackson, Ram Kugananthan, Bob Hide, James Hunt, Louise Jackson, Liliana Knesevic, Lauren Lee, Colin McKenzie, Eleanor McKenzie, Amanda Morrison, Peter Morrison, Adam Paterson, Debbie Pepler, Val Pepper, Cate Quinn, Russell Newman, Allan Robinson, Liz Robinson, Janet Scott, Sophie Saussier, Kelvin Smith, Fiona Trevett, Rachel Winstone, Roger Wood, Bill Zuurbier and the Chief: the late Terry McKevitt.

And to: Simon Adams, Jo Ashton, Rick van Barneveld, Chris Deane, Steve Down, Derek Drysdale, Jeremy Harrison, Gil Howarth, David Meek, Rory O'Neill, Peter Richards, Richard Spoors, Jim Sydall, Richard Wade and Tony Wager for their support.

And to Per Brodèn, Julian Downes and Peter Harnett for their ideas and year-in year-out encouragement, and to William Downes for his steadfast editing.

For J.

Project Risk Analysis

Techniques for Forecasting Funding Requirements, Costs and Timescales

DEREK SALKELD

Routledge
Taylor & Francis Group

LONDON AND NEW YORK

First published in paperback 2024

First published 2013 by Gower

4 Park Square, Milton Park, Abingdon, Oxon OX14 4RN
605 Third Avenue, New York, NY 10158

First issued in hardback 2019

Routledge is an imprint of the Taylor & Francis Group, an informa business

© 2013, 2019, 2024 Derek Salkeld

Publisher's Note
The publisher has gone to great lengths to ensure the quality of this reprint but points out that some imperfections in the original copies may be apparent.

British Library Cataloguing in Publication Data
Salkeld, Derek.
  Project risk analysis : techniques for forecasting funding
  requirements, costs and timescales.
  1. Risk assessment.
  I. Title
  658.1'55-dc23

ISBN: 978-1-4094-4497-8 (ebk – PDF)
ISBN: 978-1-4094-7237-7 (ebk – ePUB)

Library of Congress Cataloging-in-Publication Data
Salkeld, Derek.
  Project risk analysis : techniques for forecasting funding requirements,
  costs, and timescales / by Derek Salkeld.
    p. cm.
  Includes bibliographical references and index.
  ISBN 978-0-566-09186-5 (hardback) -- ISBN 978-1-4094-4497-8
  (ebook) 1. Risk assessment. 2. Project management. I. Title.
  HD61.S235 2011
  658.15'5--dc23

                                                          2012004312

ISBN: 978-0-566-09186-5 (hbk)
ISBN: 978-1-03-283833-5 (pbk)
ISBN: 978-1-315-60247-9 (ebk)

DOI: 10.4324/9781315602479

Contents

List of Figures

List of Tables

Introduction

About this Book

Public and private institutions have always had to make decisions about whether or not to invest in projects intended to provide society with new goods and services. Their predominant concerns will be how long a project will take to complete and how much it will cost. Their decisions will depend to a significant extent on the degrees of confidence they have in the accuracy of each of these two parameters, accuracy that project development works hard to increase. I suspect this lack of confidence comes from the persistent tendency for projects to overspend and overrun, and that it is a caution nurtured by experience.

In this book I want to make a case that the reason projects overspend and overrun is that they were underestimated in the first place. This may be obvious after the event, but why do such mistakes occur? It is this argument that I want to set out in this book. Further, so that I do not succumb to what I once heard described as the English Disease: clearly identifying the problem yet doing nothing about it. I will also propose a solution to the problem of underestimation, which I will support with worked examples and suggestions of ways to put these into practice.

The entirety of this exposition may constitute a new way that costs and timescales of projects should be predicted, and how they should be managed afterwards. The usual strategy for avoidance of a future overspend or overrun when a project is in the early stages of development is to seek ever-increasing assurance about the validity of the numbers used in the estimates. In my experience, this assurance is usually based on the closeness of the quantifications to those for similar but completed projects that have been adjusted in rational ways thought to allow for the differences between the projects: allowances for changes in the cost of labour or for changes in legislation and so on. However, every project is unique either in the nature of its work or in the context in which it is carried out. Repeatability of the cost and timescales characteristic of the production of goods is unlikely. To borrow a phrase, past performance is no guarantee of future performance. Historical costs and timescales are therefore only indicative of those that will be incurred on a forthcoming project.

Though hindsight has deemed them inappropriate, historical costs may have been the only numbers available prior to the starting of a project. I am going to challenge this because I think the *way* they were used may also have been inappropriate. Though traditional estimation methods for cost and time may be good for assessing the project work, they are inadequate for assessing the project risk, and risk is what changes most from project to project. Though the designs and specifications of a past project may be similar to a future asset, its costs and timescales can only provide limited assurance to a future project. Risk, which in the context of this book is the potential for a project to overspend and overrun, has to be funded by the project financiers, just like expenditure on labour and materials does.

If a risk analysis is not carried out then the costs and timescales of future projects are likely to be underestimates of the quantities needed. Consequently, to sanction a project on estimates alone is to make a commitment to a future overspend or overrun.

With so many types of risk and risk situations I am not making a case that what this book contains is the true article – an exclusion of all other forms. Instead I want to explain how I use risk analysis to calculate how much time and money a project will require in order not to overspend and overrun. As a result, investment decision makers can have a higher level of assurance that what they see in the proposals for a project will be reflected in both the financial and practical outcomes.

I want to show managers and analysts how they can use risk-analysis techniques to discover the actual funding required for what I hope will be a successful project. I have tried to make my explanations descriptive and non-mathematical so that the insights the methods will bring are simple to understand.

The Structure of the Book

Chapter 1 examines why risk analysis is important to an understanding of how much money and time is needed to adequately fund a project.

Chapters 2, 3 and 4 are concerned with evaluating the cost and timescales of a project, taking into account the risks for the project-funding community. Proposals for projects are usually developed by teams with a wide range of skills and experience. I hope the descriptive approach used will allow the assimilation of the tools and techniques described by all who may be involved.

Chapter 5 looks at the allocation of risk ownership. The process of deciding ownership sits at the transition point between risk analysis and risk management. Before allocation of ownership, the analyst helps and informs the funders in their investment decisions, and afterwards the project managers in their management actions.

In Chapter 6 I will explain how an analysis-led approach to risk management can provide the manager with the assurance that the project will be delivered on time and within budget, something the funder will almost certainly be seeking. It is, after all, the funder's money the project manager will be spending, not their own.

Chapter 7 looks at suggestions for an analysis-led risk management process, and some suggestions for how to put into practice. Chapter 8 reprises the arguments in the book in a conclusion. Appendix 1 provides a set of checklists for the identification of risk, and Appendix 2 contains the example risk model used extensively throughout.

Bideford, England December 2011

1 *The Case for Risk Analysis*

Why is so Little Known about Project Costs and Timescales?

In 2008, the British economy turned over just under £1,450 billion.[1] The measure of this, the gross domestic product (GDP), comes in three forms,[2] one of which is the value of the goods and services its people produce (Figure 1.1).

All of our efforts to keep the wolf from the door are in there somewhere, but when I ask people in which part of this spectrum they work, it quickly becomes apparent that there are quite a few people for whom the answer is a 'kind of neither, probably both' portmanteau expression that German speakers would have little problem expressing in a single word. They do not work in manufacturing, but similarly they do provide relief to those in need. They are neither producers nor providers of service. Instead, they say they produce a good, or perhaps just a small number of them, a specific product to a specific client. They say they work in teams that deliver motorways, bridges, computer systems, production lines, power grids, buildings and so on. They work in what I suggest is a hitherto unsung sector of the economy called 'Projects' (Figure 1.2). How big this sector of the economy is I do not know, but I would not be surprised to learn that it is both large, and that a small number (tens of percent) of the UK workforce consider themselves to be part of it.

Since working on projects is economically productive, there must be a band in this spectrum equal to the value of the projects completed annually. Where it sits, towards the goods or services end of the GDP spectrum, is not clear, but it has to be in there somewhere.

It almost goes without saying that the value of the goods the UK produces has declined in recent years. However, since the GDP has grown, it is reasonable to assume that the value of UK services must have increased

Among the goods we can no longer buy are Raleigh bicycles (started in 1887 and closed in 2003), Rover cars (1885 to 1967, when the company became part of British Leyland), and Norton motorbikes (1898 to 1975). Among the services that have replaced them are English-language entertainments (BBC TV, the West End), tourism (easyJet) and finance (providers' interests, first pensions, money invested in money, bets placed against the economy, private equity purchases of successful companies, etc.). Indeed, we may have been too successful at the last. Though the profits made by these companies may have generated taxes for the Exchequer to expend (or invest of course), being privately owned means those profits have not been accessible to the masses due to the equity no

---

1   http://www.hm-treasury.gov.uk/d/pbr09_annexa.pdf Table A9, Money gross domestic product at market prices.

2   http://www.statistics.gov.uk/CCI/nugget.asp?ID=56

Figure 1.1   The UK gross domestic product spectrum I

Figure 1.2   The UK gross domestic product spectrum II

longer being publically traded on the stock exchange. I digress, but perhaps even for a capitalist like me, life is a progression from the political left to the political right and then slightly back again.

Somewhere in the spectrum of goods and services lies the projects sector. Projects have provided most of the objects and products around us that we would consider evidence of a modern, civilised society. Many of these such as roads, railways and airports are clear, apparent to the masses, while others such as broadband, air traffic control and customer information databases are less so.

Projects are distinguishable from goods in that generally each project produces only one good before it is closed, and even though some projects are commissioned to set up facilities to produce goods, once the products start to roll out of a new production facility, the project is finished. A subsequent requirement for a further good is usually the cause for a further project.

Projects are also distinguishable from services. The idea behind a services-based business model is to reduce costs and improve quality through economies of scale and repetition. If I can work out a way of doing something you do but for which you have neither the time nor the inclination, and if I can persuade you it would be better value, then I ought to be able sell that service repeatedly to you (and to others). I would make my profit from multiple sales, probably accumulating it bit by bit as the business year rolls by.

If what you need however is a project, then you are probably only going to commission one from me. I need to make my annual profit, or rather, earn your contribution to it, from a single sale, normally taking it at the end of our business deal when the project is finished. Moreover, when it comes to building the next thing you want – power distribution network, ship, or satellite – I will probably have to persuade you to use me all over again, usually through some form of competitive tender. It would be very unwise of me to base my business plan on a presumption of a number of multiple sales to you previously.

Of course there are many companies that deliver major projects throughout their business year, receiving repeat business (without competition) from satisfied clients. Goods and services are generally typified by lots of smaller projects and sales.

I may have given an impression that projects is the harder business sector to be a part of, but this is not entirely the case. There are some compensatory aspects that are extremely attractive. First, projects is arguably a much less risky business than goods or services because project costs, staff, facilities, materials and such like, are often reimbursed by the project funder within a few weeks of them being incurred by the project delivery company. Secondly, if project-delivered goods do not work as intended, remedying the situation can possibly be paid for by commissioning a further project. This sounds slightly scandalous, but my perception is that projects sometimes deliver things that are markedly different from the original designs, and that possibly the newer items may have attracted additional funding. You must also contrast this with the money-back guarantees and refunds – necessary nowadays to secure sales in the goods sector. Finally, if the project good does not provide the benefit foreseen, say perhaps the number of drivers across the toll bridge is fewer than forecast, then that is not the fault of the project-delivery company, nor is it a loss-making scenario. The project delivery company only does what it is commissioned to do and doesn't generally take responsibility for the level of use of the good.

So somewhere in the GDP spectrum of goods and services is a 'neither one nor the other' band, a group of projects that spend other peoples' money and accrues its profit on single trades, usually some time after the work has been commissioned. There is evidence that this band exists everywhere we look: telecommunications systems, water supplies, power generation and distribution, transportation, information systems. But there is no sight of it as a sector in our economic statistics (though there is evidence in the services end of the spectrum of its use). Perhaps this economic data is available to economists. It would let them quantify the parameters of the sector, and though I am not suggesting for one moment that there is a conspiracy on the part of those involved in projects not to publish this data, I can see an economic rationale for its absence. I will illustrate this using four case histories from projects that have made the front pages and the prime time broadcasts in their own countries. They are:

- The Australian Federated Health Care Records (IT system)
- RAPID, The European Heavy Logistics Aircraft (Defence)
- The Buenos Aires Urban Light Rail Transport System
- The Sino-Canadian Polar Air Traffic Control System

I spoke at a university conference recently and asked the audience of academics and students to speculate why these projects were in the news headlines. I suspect the students took their lead from their professors but regardless, they all agreed it must be because the projects were over budget and late. No one suggested that they just opened then closed. No one suggested they were white elephants, or even posited that they were associated with infamous accidents.

My presentation was entitled 'What we know and what we don't about projects', and I think that I was successful in convincing my audience that what all of us know about projects, even those who are not involved with them, is that they cost more and finish late.

I have carried out this unscientific *vox populi* experiment a number of times over the last ten years, enough for me to be confident of the outcome. With one exception I have never had any response other than that the projects must be over budget and late. The one exception was right though: none of the projects exist. I made them up and so they are not famous at all, but in the minds of the audience projects must be newsworthy because of their flaws. It is what we have come to expect.

If I had presented the names of four fake goods in my lecture and asked the audience why they thought *those* might be newsworthy, I suspect I would not find a consensus in the answers. I also suspect the same would apply to four fake services. If there was a consensus among my audience then it would probably be a deduction made on the basis that, since all successful goods and services attain such a strongly branded recognition in our shopping society, these ones must be the flops. It would be the reason they're newsworthy. Fuzzier, more nuanced business-speak might be that the goods were poorly positioned in the market, there was a lack of quality control of the service, the underlying business model was inflexible, there was a failure to adapt to emerging markets. In reality, 'over budget' and 'late' would be suitable responses. If there is an ingrained perception of failure, both financially and in relation to time, what does one do if they are working in the projects sector of the economy in order to thrive?

This is not, metaphorically speaking, a theological debate about what people believe to be the truth. Though it may not be supported by evidence, there are justifications for this public perception. In the 1997 edition of his book *The Management of Projects*,[3] Professor Peter Morris explains how he could find the final out-turn costs for only 1449 projects in the public domain and that of these, only 12 had been delivered on or below budget. I could see that number from numerous office windows in any big city. Professor Morris also describes a later analysis of 3000 projects that found a similar result.

How then do companies that deliver projects survive when there is evidence the sector fails to deliver on time and to budget? Sorry to sound cynical, but 'on time, on budget, on occasion' is not going to be a business-winning pitch for a company, no matter how bravely honest it may be. Perhaps it would be wiser, all things considered, to keep quiet about overspends and overruns in case admission of their existence may be seen as evidence of lack of grip and drive (essential prerequisites for employment as a project manager). It is vital that prospective clients are assured that past performance is not a guide to the future.

It is also prudent for those that deliver the project not to let their performance (delivering on time and to budget) be subject to analysis by economists. Those researchers would need to identify their sources if their work were to have credibility when peer reviewed for publication. This is obviously not the case with goods and services, both of which offer researchers rich seams of data on supply, demand and pricing to mine for econometric understanding and insight.

I suspect the reason people and companies involved in the projects business survive overspends and overruns is the same as the reason why projects does not feature in the GDP spectrum of goods and services – information regarding the frequency and extent

---

3   Morris, *The Management of Projects*, (1994). Published in London by Thomas Telford, ISBN 0 7277 1693 X

of failures to deliver on time and budget is never released. Though not completely true, published data about the actual timescales and eventual costs of projects are scarce.[4] The projects sector keeps quiet and gives out nothing by which it may be delineated and measured. One is therefore led to conclude that it is not in the economic interests of any one whose livelihood is earned in the projects sector to publish information about their performance. There is no commercial advantage to be gained from doing so, and I suggest this is why the sector is not recognised in the GDP spectrum.

The Case for a Better Understanding of the Projects Sector of the Economy

Delivering projects as a business venture, in which someone provides a good to someone else for a fee, has probably been around since public works began. A Eurocentric view would choose the forums of the Mediterranean city states or the pyramids of Egypt as early example. This led to the introduction of grids of one form or another: roads, canals, rail, mass housing and so on, upon which our current daily lives have very much come to depend for their relative ease and comfort.

I estimate that approximately 250 years' worth of projects have been designed and built by project delivery businesses. Inevitably, a Web search of project management companies finds thousands, so I started to draw up a list of companies I know of that are currently practicing in the sector and looked up when they started: 1920s, 1880s, 1944, 1978, 1848, 1901, 1885. Clearly this is a large and a persistent business sector. The questions that then arose were:

- If these the companies have all been around for a long time, how could they have avoided overspending and overrunning?
- If they have all been trading for longer time frames than any of the projects they have worked on, then in spite of probably enduring their share of overspends and overruns, they must have been able to continue efficiently trading to attract further business.

My suggestion in relation to the first point is that they keep quiet about it. To the second – that like railroading in the nineteenth-century United States – dominant positions for the design and delivery of certain kinds of projects may be capable of being established through company size, experience, expertise, connections to political power and an ability to provide the funding required. One can see that access to these would make a compelling case for the appointment of a project delivery company, even if its track record of delivering on time and to budget were no better than any of the others.

Where the market demand will be for the supply of project delivery services, and what types of projects will be commissioned therein over the coming years, I do not know. The projects sector of an economy is not as well understood in terms of the benefits or encumbrance it brings to societal development as it should be. Certain projects in

---

4   Wishing to progress some research, I recently asked HM Treasury if it had any project data. It did not. I also wrote the Department for Business, Innovation and Skills to ask if it had. It did not either.

certain circumstances may be worth supporting politically in order to stimulate and develop expertise that could then bring economic benefits to the country through its export. Not the projects of science fiction: monorails everywhere and hologram holiday experiences, but the deliberate nurturing of a capacity and an ability to deliver reliable, useable and appreciated assets in markets that needed them. If you want a good car, turn to the Japanese who have plenty to choose from. But if you want, say, a potable water network or an electricity distribution grid then why not turn to the British? However, the establishment of a good reputation for delivery of projects in a particular field will always be difficult – impossible – if they are persistently delivered over budget and late. The UK has already exported 'lines and a ball' style games – cricket, football etc., and with systems of tolerable government around the world and I can see no reason why the world demand for projects should not automatically turn to UK PLC. One has only to look at the presence of US project management companies in the UK, doing what I thought we could do perfectly well for ourselves, to deduce that there is an export market for project delivery expertise. It does not matter what the types of projects are, but it does matter that we avoid overspending and overrunning.

Some Attempts to Compensate for Potential Overspends

The political interest in projects is considerable because government departments and agencies are major funders of them. As a result there have been several initiatives to develop processes that will avoid potential cost overruns on projects.

Perhaps the project with the highest public profile is the Private Finance Initiative (PFI) in which the public sector contracts the private sector to deliver an asset it then pays to use. The private sector designs and constructs, and often operates, the asset using its own sources of funding. As a result, some form of contractual arrangement is put in place to ensure the private sector accrues enough usage charge over time to cover its costs and make a profit. One key reason for procuring projects in this way has been a belief that the private sector will be more successful at delivering the projects on time and to budget than the equivalent public sector organisations.

Since the private sector tends to borrow the funds it needs to design and construct the asset, there is an argument that because the government could borrow those funds at a lower rate of interest than the private sector, its own project teams could deliver the project for less. This argument is predicated on an assumption that the public sector has project delivery organisations similar to those at the private sector's disposal. However, one has only to read the daily government procurement notices over a period of time to become aware that this may not be the case, for what was once done in-house by the public sector must now be contracted out because the public sector no longer has the wherewithal. To illustrate the point in an unscientific way, here is a typical procurement notice (Table 1.1) from the *Official Journal of the European Union* for the kind of project management services the local authority concerned, whose name I have redacted, would at one time have had the capability to do for itself (own in-house resources). Now the capability and capacity has to be bought in.

Local authorities once had departments who did the kind of work in the notice, and by inference, the requisite skill base was maintained in-house. Now that it is contracted

Table 1.1   A procurement notice for the contracting out of project management services from the public to the private sector

Notice Details	
Title	UK: business and management consultancy and related services
Name of Awarding Authority:	xxxxxxx Borough Council
Award Criteria:	The most economically advantageous tender
Type of Awarding Authority:	Local authorities
CPV Product Code:	45210000 (Building construction work.) 45214000 (Construction work for buildings relating to education and research.) 45314000 (Installation of telecommunications equipment.) 50700000 (Repair and maintenance services of building installations.) 51610000 (Installation services of computers and information-processing equipment.) 70000000 (Real estate services.) 71000000 (Architectural, construction, engineering and inspection services.) 71240000 (Architectural, engineering and planning services.) 71315000 (Building services.) 71541000 (Construction project management services.) 72222300 (Information technology services.) 72500000 (Computer-related services.) 72514300 (Facilities management services for computer systems maintenance.)
Nature of Contract:	Service contract
Type of Procedure:	Competitive Dialogue
Publication Date:	28/12/2010

out, however, the public sector expertise is in the buying of the skill, rather than the practice of the skill itself.

There is nothing in the strategy of contracting-out project delivery services that specifically mitigates overspending and overrunning, other than the assumption that the private sector can manage projects better than the public sector. I am not sure about this, because the work of delivering projects is the same be it done in the public or private sector, and I am certain there are (or were) just as many talented project people in the public sector as in the private. The public sector is at least as able as any other employer to attract project management talent, not least of which is its ability to offer what I think must be a deeply attractive prospect: using a career to build an entire system, a city say, instead of being a peripatetic builder of, for example, office blocks. The rebuilding of places like London and Plymouth in the 1950s and 60s, and of building Milton Keynes in the 1970s must have made for many a satisfying public sector career in projects.

The reason project delivery services are gravitating from the public to the private sector may be that when projects in the private sector are over budget and late this no longer emerges into the public record, but instead into a private one. Those who see it emerging may well take a commercial view that perhaps with some judicious renegotiations of the loans to reduce the interest rate, or some astute adjustment of operational and maintenance regime, any overspends and overruns can be quietly dealt with over time, well away from the public gaze and inquiries by a fourth estate.

Not all projects naturally cleave to a PFI model. Some are simply too complicated. They contain too many systems that need to be integrated flawlessly and sustainably for the delivered asset to work reliably and consistently to secure the initial funding required. A high-speed motorway project with interfaces to slower major roads may be possible to fund through PFI, but I doubt a high-speed railway with interfaces to slower lines could be. There would be too many component systems: rail, rolling stock, traction power, signalling, telecommunications and civil engineering structures. It would not be impossible of course – but it would probably be a scheme that would be too difficult to arrange, hindered by the complexities of resolving liability for train delays and cancellations among individual system providers. Warships and other military system projects come into the same category, with the additional complication of the possibility of the delivered asset to be lost when in use.

Other projects will have too many external parties whose competing needs have to be accommodated, moderated or confronted in order for a private sector project company to be able to get on with the job without disruption. The design and building of a modern Olympic Games is an example of this.

And so, though the public sector has contracted out a lot of project delivery to the private sector, in the end not all projects are suitable for such an approach. In these cases the public sector has little option but to make a presumption that the funding and timescale estimates are sufficient to get on with the project itself.

Unfortunately there is evidence of concern in central government that its devolved agencies and branches frequently fail to get these sums right. In 2002, Her Majesty's Treasury received a report entitled *A review of large public procurement in the UK*, prepared by the consulting engineers Mott MacDonald.[5] It 'demonstrates the existing high level of optimism bias in project estimates arising from underestimating project costs and duration or overestimating project benefits'. Mott analysts were given data regarding overspends on 50 projects. They split them into five categories and reached the not entirely statistically robust conclusion that projects of certain types tended to become overspent by certain percentages. These overspends were attributed to an optimism in the original estimates prepared by the project sponsors: thus the term 'optimism bias' (OB) has become a quite well-known percentage mark-up, so much so that is added to many proposals submitted to the Government for funding (Table 1.2).

---

5   *A Review of Large Public Procurement in the UK*, Mott MacDonald (2002), available at www.hm-treasury.gov.uk/ greenbook.

Table 1.2   Optimism bias percentage adjustments for duration and capital expenditure

Project type	Duration		Capital expenditure	
	Upper %	Lower %	Upper %	Lower %
Non-standard buildings	39	2	51	4
Standard buildings	4	1	24	2
Non-standard civil engineering	25	3	66	6
Standard civil engineering	20	1	44	3
Equipment/development	54	10	200	10

One may wince at the confidence with which the results of this analysis were presented, but no one really doubted the basic truth that overspends were a persistent phenomenon, and certainly no one would bet their careers and investment funds that the next project along would be underspent and delivered early. However, the data set was statistically too small to support the inferences made. Such is the reticence that surrounds the public disclosure of project accounts, it may well have been all the data there was.

In addition to the limited data set, there is another problem. I was once commissioned to calculate the size of the optimism bias mark-up that needed to be added to a hybrid project comprising a mix of civil engineering and IT works. A colleague[6] pointed out that whereas the final cost of a project should be just the one, and its value known, the other side of the ratio, the initial cost, could have had several candidates, each reflecting the different scopes, specifications, baselines and development maturities that are the typical outputs of the early stages of projects. This raised the question: in the data set used to fix the optimism bias, what had been the maturity of the initial estimates? And, to take it a step further, after the initial estimate, had the projects been successfully or badly managed? These important measures of the appropriateness of applying OB to other projects were simply not available.

I might easily have applied an OB adjustment to a very mature and well-understood project intended for delivery by an experienced team that had been derived from a small set of estimates from a number of poorly managed, overspent projects and their early days, back of a fag packet, estimations. The opposite may equally have been the case: an OB adjustment based on well-performing projects could be applied to one that was going to perform badly.

The potential for the inefficient use of funds is obvious. OB gave project managers money but the reward came with no reason why it was the amount it was. Optimism bias was meant to cover everything without explaining why and what for. It further struck me that if OB mark-ups were always applied to project estimates then it would not be long before design and contracting companies would quickly see them as an opportunity to increase their prices. Like C. Northcote-Parkinson's Law that work expands to fill the time available, prices would rise to absorb the additional funds that OB brought. Furthermore, without any communication from the past to the present about what the underlying

---

6   I am grateful to Dr Julian Downes for this insight.

difficulties were for which the treatments are an OB mark-up, I would not be surprised to find that after a few rounds of funding projects with OB applied, projects were still overspending and overrunning because managers would not be obliged to spend OB funds mitigating those difficulties. The funds may simply be expended on the works generally, for example, by the acceptance of higher tender prices. It is very rare for a project that has stepped out into the public spotlight of issuing invitations to tender to be stopped because the quotations received are slightly higher than estimated. One can imagine commercial belts being loosened all round with sighs of relief on learning that the calculations of the funding required for a project include a non-specific OB mark-up. Overall OB provides a layer of financial cover, whether or not any underlying problem is likely to pertain to a particular project or not. This may not only be inefficient, it may also be ineffective because the funds are not aimed at anything specific. OB mark-ups can also be very high for projects that must use a mix of engineering technologies, sufficient perhaps to overturn the case for doing them if they were blindly applied without consideration of appropriateness.

A more sophisticated approach was needed so the Department for Transport commissioned a more statistically valid analysis from Bent Flyvbjerg, then working in the Department of Development and Planning at the University of Aalborg, Denmark.

Professor Flyvbjerg, in association with the consultants COWI, reported previous work in which 260 transport projects had been studied.[7] Professor Flyvbjerg noticed and described a behavioural phenomenon in which people working on projects, their roles ranging from design and construction staff to commissioning agencies and funders, and even politicians at both local and national level, seem to collude unconsciously in an underestimation of the costs and timescales needed by a project, as well as an overestimation of the benefits that will be obtained once it has been completed.

There is no suggestion that there are conspiracies to get a project approved and started, or even connivances based on a nod, a wink and a tap of the finger on the side of the nose. But the sense of there being an unspoken understanding between those whose livelihoods and place in history may depend on a project going ahead can be inferred from the very readable book Professor Flyvbjerg wrote about his researches.[8]

The mathematician John Forbes Nash described an eponymous equilibrium in which apparently competing players in a market were discovered not to be doing so, each of them coming to a conclusion that the rewards they were getting were sufficient and that competing against each other for more was neither worth the risk of losing market share nor the additional effort required. Perhaps something similar exists in the project world between the stakeholders? Everyone is accruing something they value from the current phase of the project development, and they subconsciously understand that if they behave in ways that fulfil the assumptions and needs of the others, then the next stage of the project will more likely gain authorisation to proceed. Consequently, they may stand a decent chance of continuing to be party to it and thus sustained in business. All it may take for a project to continue in development is that no one should rock the boat unnecessarily, and one way to do that would be not to show that the costs and

7   The British Department for Transport (sic). *Procedures for Dealing with Optimism Bias in Transport Planning*. Report No. 58924, June 2004.

8   Bent Flyvbjerg, Bjorn Azelius, and Werner Rothengatter *The Anatomy of Major Projects*, Cambridge University Press, 2003. ISBN 0 521 00946 4.

timescales are intolerable but instead to suggest that with further diligent research, a way to make them acceptable will be found.

Professor Flyvbjerg improved on the Mott MacDonald analysis by devising a percentage mark-up that should be added to the estimated costs and timescales for a project if it were to have a given level of confidence in the final costs and completion time scales before it was authorised to start (see Footnote 8). The greater the level of confidence required, the greater the mark-up to be applied. The curve in Figure 1.3 shows the percentage uplift that should have been added to the estimates for projects in Professor Flyvbjerg's data to deliver roads for an acceptable chance of a cost overrun.

From the figure, if only a 10 per cent chance of overrun is tolerable, i.e. there must be a 90 per cent chance a road project must be delivered within budget, then the estimate must be increased by approximately 65 per cent.

Though this was a far more robust analysis, and one of considerable value to financial analysts, it is still susceptible to the same drawbacks as the Mott McDonald one when it comes to commercial and managerial practicalities: the mark-up does not cover anything specific so there is no need to spend it on the treatment of the causes of overspends and overruns. Whether it would be better to spend the money on hedging against demand-led cost increases or setting up a systems integration and test facility separate from the main operational facility is not known. There is simply a pot of money waiting to be used for something unspecific that may or may not emerge. Indeed, because the mark-up is derived from an analysis of many projects, when it comes to a particular case of a project in difficulty it may be more than sufficient. Of course it may also be insufficient, and so the potential for inefficient and ineffective use of funds remains. Further, given a non-specific fund to expend, how should the project manager use it? Do they keep the fund as a contingency measure, for use in the event of something deleterious? Or should it be used to punt on a pre-emptive act intended to avoid it? If the former and the event happens, how can the manager be confident it is not the first of many, and thus able to judge how much of the fund to use and how much to reserve?

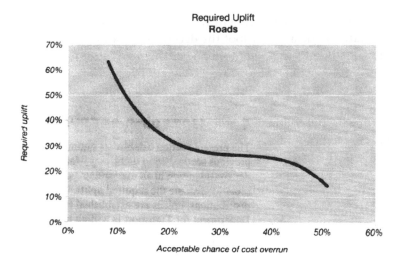

Figure 1.3   A Flyvbjerg curve

I think the efforts to evaluate a compensating mark-up to cover project overspends are rational, a good thing and academically sound, but a mistake from the standpoints of project funding and management because of the possibility of the inefficient and ineffective use of funds and because managers are given little intelligence about the what, why and when of overspends that they can assimilate and perhaps mitigate.

Projects overspend and overrun. They have done so enough for academics to have noted it and for the UK government to have tried to do something about it: but the solutions I think are unsatisfactory. I want to propose a better one.

It may not, in its early stages of development, be any more accurate (in its predictions of outturn costs and timescales) than the mark-up solutions, but I hope that its practice will stimulate challenge and improvement so that ultimately it is seen as a much more accurate and insightful approach.

I suggest projects overspend and overrun because they were not estimated correctly in the first place. If I were a scientist seeking to advance human knowledge and I devised a scientific hypothesis that was subsequently invalidated, I would try to think of a better one, one I hope would not be confounded by the data. It seems to me that the experimental evidence tells us that projects cost more and take longer than intended, and accordingly that this must invalidate the ideas, and hence the methods used to estimate their costs and timescales. I don't think the calculations used are entirely the right ones. In the same way that Newtonian mechanics begins to fail experimental tests at the subatomic level, resulting in the need for quantum mechanics, I suggest the basic methods of estimating the money and time needed for a project begin to fail when a project becomes big in some way or ways: physical size, number of components, geographical spread, length of time needed etc.

What I believe would make the estimating methods more accurate is risk analysis.

An Approach to Forecasting the Potential Overspends in Project Costs

I would like to describe a small experiment which, like the fake projects I listed earlier, is a bit of a cheat. Mathematicians will spot the deception in an instant, but even so, it has worked so well with students that I am told one of them has adopted it for use in pitches to potential clients. I hope you will tolerate its naivety because a point needs to be made.

Imagine I want to commission a project to build a garage with a drive leading up to it. I approach a builder and he gives me a price of £3 for the garage. The numbers are implausible but make what follows easier to assimilate. The builder only erects structures and so I have to hire a ground works contractor to lay the drive. He too gives me a price of £3. I then apply to the bank for a loan of £6 to fund my project.

The bank are concerned about this scheme and call in a risk analyst (it is a fantasy story) who then researches the supply and demand of garage components in the marketplace, as well as what can go wrong with their assembly and commissioning. They find that the supply of components outweighs the demand. As a result they can be readily purchased for £2. Some suppliers even want to offload stock and are prepared to sell them for £1.

Research into problems with existing garages finds that if the slope to the base on which they have been erected is not level, it can cause the embrasures for the door and window openings to be out of true, necessitating the planing of the doors and windows to fit the hole. These extra works can add up to another £3 of costs for the garage.

We now have a scenario in which the garage could cost between £1 and £6, the former having no problems, the latter a full set of commissioning problems. The £3 estimate has, in risk analyst argot, a risk–opportunity spread of £1 to £6 (please ignore reversed order of the numbers vis-á-vis their description. It is easier to say, that is all).

Research also finds that the concrete has an estimated cost of £3, but demand is weak so the order may be available for £2 (or even £1 if a cancelled order for elsewhere can be taken). The quantity needed however is variable because of undulations in the ground, and an additional smooth mix may be needed to give an acceptable finish to the top surface. Again the analyst finds a £1 to £6 risk–opportunity spread on the £3 drive.

The problem is how to add that up? Adding £3 for the garage and £3 for the drive is easy enough, but £1 to £6 added to £1 to £6 is counter-intuitive. Thankfully, it can easily be calculated. Imagine two dice from a board game, one to six; one red, one white. The red one is the cost of the garage, the white the cost of the drive. I have two here that I will roll as I write:

Throw 1:   Red 2   *White 4 … cost of the project = £2 + £4 = £6*

but that is not necessarily the cost because

Throw 2:   Red 3   *White 6 … cost of the project = £3 + £6 =£9*

which is not necessarily the cost either because

Throw 3: Red 2 White 1 … cost of the project = £1 + £2 = £3

And so on. Each throw generates a perfectly valid cost for the project. Any number on a dice is as equally likely to occur as any other number.

Some more throws:

Throw 4   *total £7*

Throw 5   *total £8*

Throw 6   *total £7*

Throw 7   *total £10*

Throw 8   *total £4*

Throw 9   *total £7*

Throw 10   *total £3*

and so on.

If these different but valid costs are plotted on a histogram to show the number of times each total has occurred, and the dice are thrown another 100 times, a pattern begins to emerge (Figure 1.4).

Obviously the minimum total cost of the project will be £2 and the maximum £12. But what also becomes apparent is that £7 occurs more often than any other total cost. I can conjecture that if I incorporate any risk–opportunity spreads I can find the estimate for a project. The most likely total cost seems to be more than the original estimate without them.[9]

If I had borrowed only £6 with which to build the garage, the most likely outcome of the project would be that I would run out of money and have to seek a further £1 to make up the £7 needed. By not researching the risks and opportunities, as well as not including those identified in the funding calculations, I would have booked myself a place among Professor Morrison's set of 1437 rather than the 12. My project would most likely be overspent in due course.

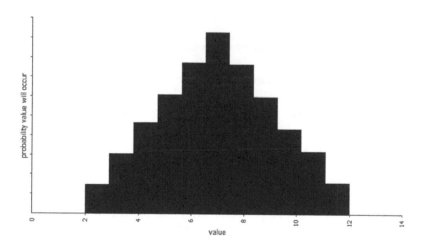

Figure 1.4   Histogram of valid project costs

I think there will be some irritation over the above triviality – I acknowledge it is a long way from a nonsense garage project to a real-world major infrastructure project, so let me close this part with five serious and worthy points that I hope my example showed.

1. When we take into account the risk and opportunities, our understanding of the funding required alters. In gaining this insight we also discovered that the risks and opportunities data underlying this result were accessible and capable of being analysed quantitatively.
2. If the risks are larger than the opportunities, and they usually are in real-world projects, then the funding required will be more than the estimate originally indicated.

---

9   In truth this is because a spread of £1 to £6 has an average value of £3.50, and it is this increase of 50p on £3 that, with two dice in the mix, gives the £1 increase from £6 to £7.

3. If we had proceeded with the project on the basis of the estimate alone, in due course we would probably have found the project overspent.
4. Not only did the calculation show the minimum, maximum and most likely cost of the project: £2, £7 and £12, it also revealed that if we were to accept a contractor's offer to deliver the garage for £6, we would know that we really ought to set aside a contingency fund of up to £6 more to accommodate the potential for final costs to increase up to £12. We now know the size of the overall financial reserve needed to cover the total of the risk exposure of the project and the estimate of its cost. In gaining this insight we also found what was needed to avoid incurring the £12 cost – make sure the base is level.
5. If we wanted to give the contractor an incentive to deliver the project for less in exchange for giving them an opportunity to increase their profit, we can see to what extent it would be reasonable to do so. Setting a target of £2 for the garage might be somewhat over-ambitious, but £4 or £5 would be fair, in exchange for a bonus of half of the £1 or £2 that may then be saved.

By taking into account the risk opportunity spreads, we come to know the minimum, the most likely and the maximum cost of the project. We also have an analysis-based idea of the size of the contingency fund we will need to cover our exposure to the risks of the costs increasing. What's more, we know the possibilities for cost savings that could be reasonably incentivised. This set of five insets has always struck me as a revelation, and moreover it is one that provides pointers to where management action can be deployed to good effect. I think the addition of the risk and opportunity analysis gives an overall result that is so much more insightful than simply adding a percentage mark-up to £6 in order to compensate for the historically optimistic estimates of garage projects.

When presented, this simple example above has all too often been immediately challenged by students who point out that there is must be more than just two risk–opportunity spreads on a real project, and that surely they are unlikely to be conveniently valued at £1 and £6. This is perfectly true.

On the first point, the number of risk–opportunity spreads on a real project is almost certainly more than two. I have never known it to be less than ten, and I am really only beginning to be confident I have an understanding of a real project when I have found at least twenty. It could indeed be any number: 10, 20, 100. The most I have seen was about 1500, a total that would prompt me to check whether or not the identification of the risks and opportunities had not been somewhat overdone and become a form of self-serving cottage industry, the sanctity of which was a heresy to protest against. Imagine then having a jug of dice, each one representing one risk–opportunity spread of £1 to £6 and repeating the experiment of throwing them, adding them up and plotting the totals (Figure 1.5).

It is easy to see that for every dice to show a one as a result of a shake and throw would be as unlikely as everyone of them showing a six; and that there would generally be as many high values (four, five, six) showing as there would be low ones (one, two, three). Therefore, we would expect the total to be somewhere in the middle, between the maximum number of ones and the maximum number of sixes. The minimum, most likely and maximum pattern of the two dice case would therefore emerge with the many-dice case, albeit stretched over a wider range but still with a peak in the middle.

And if the dice were not all one to six, but designed like those in Figure 1.6?

Figure 1.5   Many dice

Figure 1.6   Many multi-faced dice

Unless the dice are loaded, the chances of any one face on a dice being shown is the same as for any other face. The idea that all dice should show their minimum values on a single throw, or equally, their maximum values, is increasingly unlikely in proportion to the total number of dice in the bucket. The more dice there are, the less likely this is, just as is the case for a bucket full of one to six dice. The general outcome is not altered by having multi-faced dice.

However, the most likely value is not as readily apparent. It all depends on what numbers are written on the faces of the multi-faced dice. It is usually there, however, and a few hundred 'shake, throw and add-up' cycles will reveal it.

Generally, no matter what the risk–opportunity spreads are in a project, either in terms of their number or their values, when we make an attempt to measure their combined effect we will find a minimum, maximum and most likely cost. We will be able to assess the overall risk exposure and the potential for cost reduction.

If you will permit another smoothing over of the actuality, the final result looks something like this (Figure 1.7). I have fitted a smooth curve over a typical histogram and labelled it with the five pieces of information given in the previous paragraph.

Now let me put a question to you. If you were going to manage or fund a project, a new hospital say, upon which your personal reputation and well-being (your self-esteem, your choice of house, your options for your children's education, where you take your holidays and so on) depended, and you could only buy one set of information to help you secure a successful outcome to the project, which would you choose, the cost estimate or the above curve?

To me the answer is obvious: the curve. The next figure (Figure 1.8) shows a situation where the single valued estimate of the project, the spot cost, lies within the spread of the curve. To choose the spot cost as the project budget would be to take a significant chance (shown by the shaded area to the right of the figure) that at some point in the future it would become overspent. Though it is a contrived scenario, in this case the probability of an overspend would be about 75 per cent – the portion of the area under the curve to the right of the spot cost. Further, because the analysis that had gone into researching

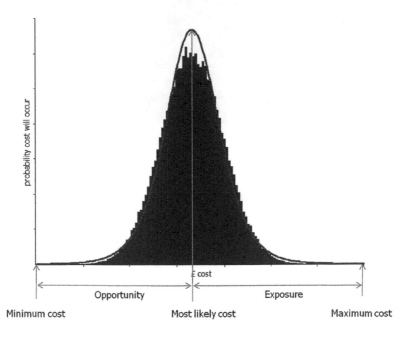

Figure 1.7   Spread cost curve

the curve would not be available to you had you chosen the spot cost analysis instead, you would probably have little evidence available to help put together a strategy to pre-empting it.

If, however, you would choose the curve over the spot cost, and were consistent in this choice over a career's worth of projects, then there is a slightly startling implication: you and I would be accepting that there is no such thing as the cost of a project. There is instead its cost function (cost curve if you like).

To be more precise, there is no such thing as the cost of a project that is not qualified by the probability of its occurrence. Presuming that your choice would be the same as mine, the curve over the spot, we will be far better informed. As a result we will know what funding we need, and just as importantly why we may need it.

Professor Flyvbjerg was right in his retrospective findings. His curves are the cost curves for different types of projects. They show the effect of the risk opportunity spreads a posteriori, which is to say after the event. I would now like to explain how to calculate the curve for any specific project a priori, in advance.

I hope I have made a case for risk analysis to be used to estimate the funding a project will need. I would now like to propose how to do it.

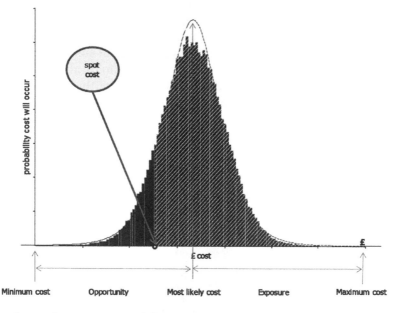

Figure 1.8   Spread cost curve with spot cost

Using Risk Analysis to Determine the Funding Needed for a Project

I think it important I propose a theory for the use of risk analysis to calculate the funding needed by a project so that it is capable of falsification and thus open to improvement or replacement.

The basic theory for estimating the cost of a project is simple and intuitively obvious to begin with. A conception of what the project has to make is broken down by the engineers and managers involved into its constituent parts, down to a practicable level usually set by the rational units of purchase of these parts. The quantity required of each is then calculated. If the component is bulk material, say sand, and the unit of purchase is cubic metres then it is the numbers of these that is totted up. If it is computer hardware then is probably circuit boards. The unit cost of each item, and the quantity required, are multiplied together to give an overall item cost. All the item costs are added up to give a total cost. So far so trivial, and I feel slightly embarrassed writing it down.

The next stages of basic theory are more difficult to apply and call for more experience and intellect in order to produce valid answers. Validation needs to be considered as something distinct from verification. The latter is assurance that the calculations were correct, whereas as validation is assurance that they were the right ones to do. They are not the same thing, and it is validity and not verification that is the issue here. First, it needs to be determined what types of labour are required, as well as how long it will take to complete such work to an acceptable standard. Secondly, what facilities and services have to be provided for them, be it an office or a production plant? Thirdly, the costs associated with the logistics of getting everything to the client's desired location need to be determined. Finally, it is a matter of commissioning, testing, training and trialling as what has been delivered is brought into service.

The steps in the basic theory of estimating are as follows:

- Specify and cost the materials.
- Specify and cost the labour.
- Specify and cost the facilities and supporting services.
- Specify and cost the logistics.
- Specify and cost the commissioning.

Experienced estimators will almost certainly feel the total cost calculated thus far may not be quite enough. They will add additional money to cover the costs of overcoming emerging adverse situations which they know their client would reasonably expect them to have foreseen, for example, that another project happening at the same time has cornered the market in a type of material. The estimators will probably have caveated their calculations with the assumptions and exclusions to forewarn the client that not everything is covered by them.

This basic theory seems intuitively right. If all the costs of all the items and services needed are known, then what else is there? The theory seems not to have worked for projects of the scale analysed by Flyvbjerg and Morris. It might work for simple projects, but it does not seem to me to scale-up awfully well to projects that are large, complex and protracted in terms of timescales. That they are paid for by one team of people but delivered by another also seems to be a persistent characteristic. Somewhere along the path of increasing project complexity, the basic theory seemingly fails. Further, no additions to the costs of most projects with published historical records would seem to have been adequate (presuming that estimators have always added mark-ups to cover unforeseen eventualities, and that the Flyvbjerg and Morris data are not the result of a fluke from projects that had no additions made to their estimates by incautious estimators, an endangered species if it not already extinct).

Something is missing from the basic theory if it is to be used to inform the funding decision and this is the risk. If we were a contractor engaged to design and build a shed, and we found a disused well when digging the footings, we would go to the sponsor of the garage project and ask for an increase in our funding to fill in the well, or bridge it, or perhaps relocate to another site. We would do this because it is not our fault the well is there. But it is not the sponsor's fault either. What is certain however is that if the sponsor wants the garage built, they must pay for something to be done about the well. In commissioning the garage project, the sponsor was therefore taking the risk that there would be no well. But what if the sponsor had looked into what might go wrong with the project by asking the neighbours or even ourselves, the contractors, about our previous experience of carrying out similar works in the locality and so had come to realise there might be wells in the area? Then it would be prudent of the sponsor to obtain not only sufficient funding needed to cover the cost of the works, but also enough to cover the risk there might be a well.

One of the attractive features of the basic theory is that to use it requires only a knowledge of arithmetic, and thus estimates can always be computed by anyone capable of determining what the other information needed may be: which materials, which labour, what facilities, the how of the logistics and the means of introducing into service.

One of the unattractive features of the expanded theory I am going to propose is that it is mathematical in nature, not arithmetical. It requires more than plus, minus, multiply and divide. It uses calculus, statistics and equations and as a result is not as simple without appropriate training. By way of developing the argument, take the economy of the UK. This is surely a much more complex thing to model in terms of numbers and formulae than the business of a corner shop. If we happened to discover that the former was modelled in Her Majesty's Treasury using nothing more than arithmetic, I think we should all be rather surprised. And so, I have always been mildly incredulous to find that the calculations used for estimating small projects no bigger than a station roof repair worth a few thousand pounds are the same as those used to estimate an entire railway requiring billions of pounds to be spent over many years. The cost of a complex project is surely not just the sum of the products, the quantities and rates of its parts? Surely it is more sophisticated? The method used to calculate the cost function for the simple garage project used not only quantities and rates, it also used formulae (the risk–opportunity spreads), statistics (the probability of things happening, which in that example was 100 per cent but could easily have been another value), calculus (the act of choosing a budget from the curve at a desired level of confidence is equivalent to mathematical integration of an area under the curve up to a boundary), and simulation (the Monte Carlo method of repeatedly putting random numbers into a set of equations that together define a model of a system in order to infer the statistical behaviour of that system – its mean, most likely and maximum costs in the garage project). This is the level of sophistication necessary to estimate complex projects.

The cost of a project is a valid concept only in the context of a completed project. It is one of those things that can only be known after the event, and this can be inferred from the careful avoidance of the use of the term by estimators everywhere in favour of the term 'estimate' when talking about projects that have yet to be started or completed. As I hope the trivial garage example in the previous section showed, what can be known about a project prior to its completion is its spread cost curve, one that has yet to start.

Hiding behind the insights of the cost function curve is a difficult decision that someone has to make: how much money ought to be allocated to a project, since there is clearly a range of possibilities with associated levels of confidence in them. Choosing a value from the curve is to decide the funding that will be made available to the project team for delivery of the project. The amount of funding is not the amount desired by a project team to deliver the project. I am sure the team would rather have something more towards the right-hand side of a curve while the funder would rather have something more to the left. The pre-project funding requirement is shown diagrammatically in Figure 1.9.

The height of the stack is the funding required for the project. It has two parts, expenditure and exposure. The expenditure is the part of the funding calculated using cost estimating techniques, and assuming a perfect world in which all of the material and labour components required for the project can be determined in advance, the expenditure is a fixed sum. A funder will probably set a budget for the manager against the expenditure. If they are generous it could be 100 per cent, though usually some is retained and so the manager is 'incentivised' by receiving, for example, only 90 per cent.

The other part of the funding is the exposure. This is the element calculated using the risk-analysis techniques seen in the garage project. Assuming an imperfect world, not all of the risks to the project will be known in advance and so the exposure is open-ended, as Figure 1.8 indicates with the increasingly fading blue shading.

The practical solution to the question of how many risks does one add into the exposure calculation is best answered by experience. Risks should be added to until the risk analyst is confident that the exposure is realistic and rationally complete. The only risks that remain to be added are those that are irrational (Mars may invade); or that are trivial matters covered by ordinary business (it may rain tomorrow); or that require incompetence on the part of competent people, processes and systems. This is not to say that people, processes and systems will never go wrong, but rather that they are not

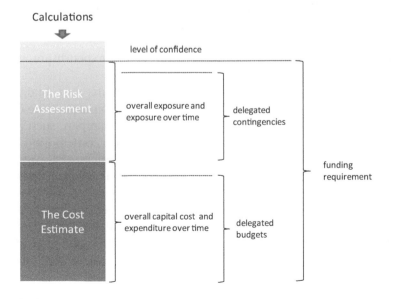

Figure 1.9   The project funding requirement stack

deliberately guided to fail. Risk analysts are therefore allowed to assume good intentions from all those who are contributing to the success of a project.

This is a different level of confidence to the one that says the funding is sufficient for the project. It is down to the analyst to make the judgement as to whether the calculations are all in there, and as such it is equivalent to an estimator deciding that the materials and labour items have all been listed in the estimate. My advice is that if the unwanted outcome of a risk has a clear possibility of materialising during the project which is not necessarily confined to the period of its design and construction, but can include the period of use, then funding needs to be provided for the risk.

Once the exposure has been computed, the funder will probably set a contingency fund at a desired level of confidence against it. This fund may be shared with the manager in some form of bottom-tier, top-tier manner, or it may be retained entirely by the funder. It depends how often the funder wishes to be bothered by the manager asking for withdrawals from the contingency fund.

Exposure to risk in practice only becomes a liability when commitment to expenditure is made – no commitment, no risk. However, the value of the exposure is independent of the value of the expenditure. For example, consider systems integration risk – design faults will emerge during the integration of diverse systems into a single entity, such that it will not work as expected. Though the liability for the risk is taken on once a decision to design and build the system is accepted, exposure to it only occurs during the later stages of the project, when the subsystems are assembled into a whole, and commissioning is attempted. The size of the exposure is low to begin with but rises later.

In contrast, expenditure on component subsystems will be higher during the early stages of the project, probably falling to zero in the integration phase. Even though both expenditure and exposure are initiated simultaneously, they vary independently of each other thereafter. Consequently, exposure is not a proportion of expenditure, even though they are often both measured in money.[10] To put it bluntly: exposure is not expenditure.

Making it so is a common error in business-case modelling, and one with potentially serious consequences. Which manager would like to discover the size of their available contingency was lower than their current exposure when problems began to emerge, say, during systems integration?

Exposure is also spread over time and has a profile like expenditure, though not one with the same shape (Figure 1.10).

Other analysts and managers will use the terms estimate, cost, budget, assessment, exposure and contingency differently. Often, risk is used to purport the contribution to the cost curve attributable to *events* that may or may not happen, such as things going wrong. Contingency meanwhile is sometimes seen as the contribution attributable to *errors* in measuring the quantities, and the *variations* in prices of things that make up the projects shopping list. Then again some will say risk is beyond what we may have analysed and modelled, that it is the 'unknown unknowns', to use the famous words of US Secretary of Defence Donald Rumsfeld. I used to get het up about others using terms I use in different ways to my own preferences, but I am more relaxed about it now. I happily take on board and use terminology that others are comfortable with and translate or interpret as is necessary. It seems wasteful (of money and energy) to debate

---

10   Expenditure is conventionally always about money. Exposure however is sometimes expressed in other currencies: lives lost, commodity shortages and so on.

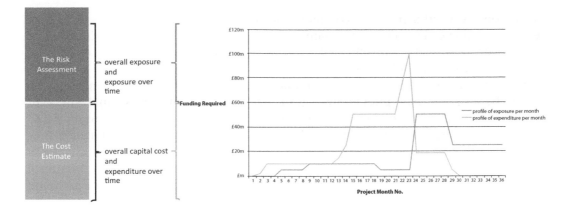

Figure 1.10 The stack with profiles

the differences, as if the business of risk is what we're being paid to manage, not the risks themselves.

I do however have one exception to this terminological laissez-faire, and that is the acceptance into the risk lexicon of 'unknown unknowns'. I doubt they exist in projects that are being planned and developed. All of the project risks I have experienced were available, even if that was only obvious with hindsight. If I had been more diligent in my inquiries, someone somewhere could have told me about it in advance. For project people to use the phrase 'unknown unknowns' in their analysis is tantamount to an implicit acceptance of incomplete work.[11]

However, I do believe that 'unknown unknowns' exist in the external context in which a project is being developed, such as emerging changes in societal or political needs that may render a project superfluous, changes in demographics that will make it a bad investment to complete, or developments in a perceived military threat that would make the deliverables produced by the project inadequate. It would be irrational of the public and their tribunes in the press and political sector to describe a project as having failed when the reason is the emergence of these bigger issues. It is unfair to expect the funders and the project delivery teams to foresee the 'unknown unknowns', and it would be wasteful to expect project budgets and timescales to hold contingencies for them just in case. To adapt the words of John Maynard Keynes, if the facts change then I change my project. Projects should therefore be expected to be cancelled or have what they set out to deliver changed when they are part-way through.

What is fair though is a presumption on the part of the funders that the delivery team are competent and experienced enough to be capable of identifying everything that could go wrong with a project this side of flights of fantasy. They have allowed for these in their calculations of the cost curve. All that could go wrong on a project can be identified by due diligence on the part of the analysts researching a cost curve. I always challenge the existence of percentage mark-ups for 'unknown unknowns' in risk

---

11   I have seen a conference presentation by Dr James Robertson, Chief Economist of the UK National Audit Office, in which he shows that every reason for an overspend or overrun identified by his office had been foreseeable from the outset of the project.

calculations, suspecting poor or incomplete research. The emergence of a Black Swan may well change the need for a project to dam a river to make a reservoir, but the discovery of porous bedrock should not invalidate its cost curve. As I am wont to say to my colleagues: there is no excuse for not knowing.

Should a funder wish to be more cautious than an exposure calculation indicates, perhaps because they have prior knowledge of changes to economic policy that will occur during the project lifetime, then it would be right for them to adjust the cost curve so that it reflects what they know but which was unknown to the analysts. Equally, they may be considering two or more scenarios, say a 10-year operating life with a change of ownership then, or a 20-year operating life closed by a decommissioning and removal stage. In these circumstances I would always produce a discrete cost curve for each and never conflate the two (Figure 1.11).

The cost estimate calculation is the summation of the products and prices needed by the project. It produces a number. The companion calculation of the risk exposure is a simulation of the overall behaviour of the risks, each of which is modelled mathematically by appropriate formulae. This calculation produces a curve. When the spot cost from the estimate is added to the curve of the assessment, the combined result is the cost curve.

Spread cost curves come in two types: with or without the estimate included. Most analysts calculate and report the latter first, then are asked to prepare and submit the former for inclusion in financial submissions to governments, public sector bodies, financiers and project sponsors. This is because it is more concise to describe the funding requirement as a single value number at a level of confidence than as an estimate and separate risk exposure.

Projects are perceived to have failed when they are over budget and late. It strikes me that a good strategy for success would be to avoid the causes of failure. To decide to fund a project on the basis of both its cost estimate and its risk assessment seems to me a good way to do this.

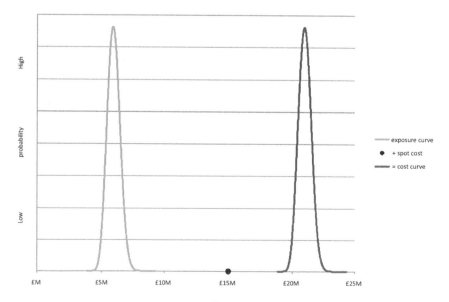

Figure 1.11 Exposure curve and spot cost giving the cost curve

2 Risk Modelling Primitives

To calculate a spread cost curve requires risks to be modelled mathematically and to do that requires their behaviours to be understood. A mathematical model of a risk should accurately match the behaviour of the risk. The set of behaviours that can be modelled needs to be sufficient to cover all possible risks.

To model a risk requires an understanding of its behaviour and a 'Lego' box of different mathematical constructions that can be assembled in appropriate ways to represent that behaviour. Thus, if in the garage project we have a risk that we may need more concrete than the dimensions of the garage indicate, because we do not yet know if a concrete path will also be needed, the analysis would of the risk would be as shown in Table 2.1.

Table 2.1   Analysis of a simple risk

Behaviour	Risk modelling elements
(A) The quantity of concrete needed might be more than the amount ordered.	1. Something that can be used to represent the possibility of the situation happening or not. 2. Something that can be used to represent the variable amount of concrete required.
(B) If the quantity of concrete is more, then additional time will be required to make the path.	3. Something that can be used to represent the additional time. 4. Something that can be used to model the financial consequences of the additional time.

I call the risk modelling elements Primitives. This term is borrowed from software engineering, where a Primitive is an element of a program language, further division of which would cause all practical significance to be lost. For example, in the IF–GOTO command of many program languages:

e.g IF X = 2 GOTO 500

500 is the line number in the programme that is executed next if X = 2. 'IF' and 'GOTO' are the primitives. Splitting them into I, F, G, O, T, O would lose them their utility.

I am going to go through a set of risk modelling primitives that can be used to correctly model the behaviour of risks. Like a computer program, risk-modelling primitives can be assembled into a risk model that can be executed in a computer to produce something of value, which in this case is not a jump to line 500, but to the cost curve.

I will explain each one briefly to give an idea of the size of the full set, what they are and how they can be used. I will then give a worked example of each in an example risk model.

One further point before we leave 'Lego' land. I indicated several primitives will be needed and would like to explain why. Given a very large bucket of Lego and the job of faithfully replicating a one-tenth size model of a cathedral, one would probably think this a futile but perfectly possible task and simply ask how much does the job pay. If on upending the very large bucket of Lego one found every piece was a grey 20 × 40 dot base plate, one would immediately think, 'hang on a moment, the model cannot be constructed from a "one shape fits all" kit of parts'. One needs a variety of pieces to represent the cathedral faithfully. As is the case with risk models: one is likely to need a range of primitives of different types in order to model all of the behaviours of all of the risks to a project in order to calculate its cost function.

The exact usage of a primitive in the example model will require several subtle variations to represent properly all the possible behaviours of risks. As a result, more primitives than are set out in the next section are actually used to calculate a cost curve. This is a basic set however. In Lego terms, the basic set is the shapes and the variations are the colours.

Primitive 1 Spread on Quantities

The primitive is used in risk models where the risk is that there is an uncertainty in the quantity of something: materials, labour, services, anything. Figure 2.1 illustrates four common examples.

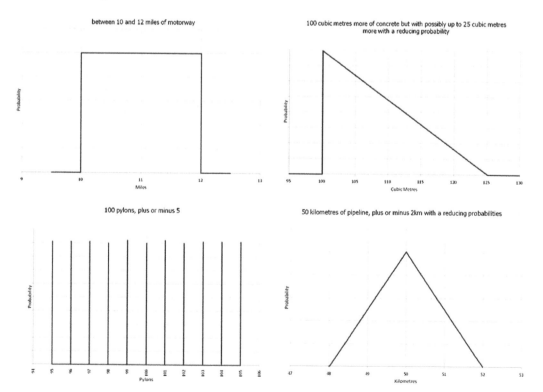

Figure 2.1   Spread on quantity: some common forms

These are estimating tolerances, and some would say such situations are not risks because there are no probabilities associated with them. However, it is clear that risks are present. I would argue that if I were to ask the funders of a project, the people whose money is about to be spent, to consider the possibility of five more pylons being required, or up to two extra miles of motorway being needed, and if they considered this to be risk, they would say yes. These are future situations where, potentially, the amount of money they have injected into the project could be overspent. Equally, we may find there could be five fewer pylons and a saving of two miles of motorway, an opportunity for the spending on the project to be less than has been provided.

Often, the argument that estimating tolerance is or is not a risk is lobbed back and forth. If I have not put such scenarios into the risk model, I will not calculate the correct cost curve. I will mathematically model this primitive in the next chapter.

Primitive 2 Spread on Price Per Item

This primitive is used in risk models where the risk is the uncertainty in the price of something a project needs to purchase: materials, labour, services, etc.:see the examples in Figure 2.2.

The same argument about whether this is a risk or not, set out in the previous section 'Spread on quantity', pertains here, as does the same response: if it is needed in order to calculate the cost curve correctly, it goes in.

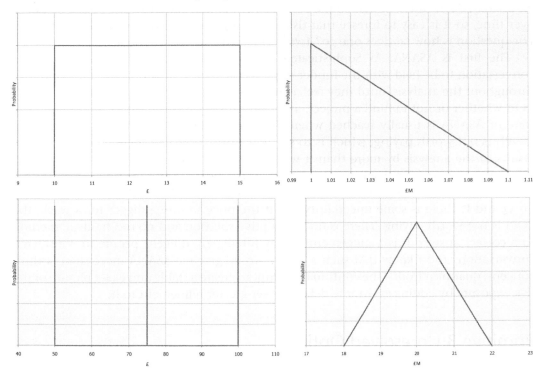

Figure 2.2   Spread on price per item: some common forms

I mentioned that some subtleties would emerge as necessary adaptations to the primitives for their correct application in risk models. To show what I mean, here is one. I will use the term price in what follows, and not cost, adopting the convention that a price is what we pay and a cost is what it cost somebody else to provide it for us. Suppose we discover that we have to model a variation in both quantity and the price of the same thing, we are saying that both might turn out to be different than the spot values used in the estimate.

However, the quantity and price may be related in some way and not free, or independent, of each other. But they may be interdependent. For example, if the quantity turned out to be more, a bulk purchase, a discount may be given, thus the price paid per item would be less. Here, the quantity and price move together but in opposite directions. The more we buy, the less we pay. The quantity and price have an opposing phase.

On the other hand, the item might be scarce and an increase in the quantity might stimulate an increase in the price. Here the quantity and price move together in the same direction. The more we buy, the more we pay. The quantity and the price have the same phase.

We have to decide which of these: independent, interdependent in phase, or interdependent in opposing phase, is the correct behaviour of quantity and price, and in the modelling ensure that the primitives that reflect this behaviour most are appropriately used.

The behaviour could be even more subtle; the item might be bought over a period of time, as and when it is needed. As a result, the price we will have to pay is determined on the day, determined by demand. Consequently, we may have to model price variation over time, so it is easy to foresee that risk modelling can quickly become sophisticated. The question is how far does one go? Two connected concepts can help.

The first is ASANA: As Sophisticated As is Necessary to Assure. This requires a conversation to be held with the funder or project manager, possibly one that continues throughout the analysis, until they feel assured that they can make a calculated decision as to whether or not to fund a project, and if the answer is yes, with what amount. The state of ASANA is usually reached when the second, the Price Of Perfect Information (POPI) is not worth paying, which is to say that further research is unlikely to alter the results of the analysis by more than it will cost to conduct it. If a cost curve produced thus far does not contain spreads on quantities and prices then I would suggest ASANA is not achieved and the POPI is worth paying. On the other hand, if within the spreads on Q and P, there is some uncertainty about the price to hire a digger for a week, the POPI is not worth paying. There is no prescriptive solution and no mechanistic method to a correct answer. It is a judgement call by intelligent, diligent people in considered conversation, and I know that such a debate would give me ASANA more than, a third concept, quantum froth – the continual attempt to create a stable and exact picture of the risk exposure by adding ever-more detail in ever greater refinement to it.

Primitive 3 Unresolved Option

Implicit in the calculation of a cost function is a presumption that construction of the works of the project has not yet started. They may not have even been fully designed, and indeed some of the early funding is likely to be the evaluation of concepts and of strategies for their construction. Risk modelling is essentially a pre-contract activity, that is to say, it takes place before major construction contracts have been awarded, and while funding is still being sought. There are other important risk tasks that are performed post-contract as well, i.e. after major construction contracts have been placed.

This primitive is used when the risk is that designs or construction methods have not been decided, and so there are some unresolved options about what is to be delivered or how it will be done.

For example, an IT project might have not yet decided whether to use Microsoft or Linux; or an electrical power project, which route to take between the supply point and the distribution panel. These are unresolved options on design (Figure 2.3).

There could also be unresolved options about whether to install the software during one big shutdown or as a rolling programme, user by user, and the cable route for the power project could be constructed from station to panel or from panel to station according to the availability of access to the route at different times. These are unresolved construction options (Figure 2.4).

The problem that the cost estimator has is that they must make an assumption about which option will be chosen to do their calculations and so have to caveat the answer accordingly. That other outcomes may be possible is potentially a risk to project funding so these have to be incorporated into the risk assessment.

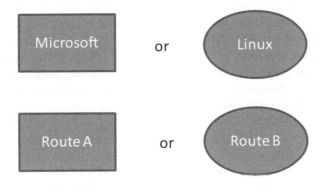

Figure 2.3   Unresolved options on design

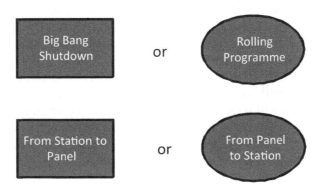

Figure 2.4   Unresolved options on construction

There is a view that this type of situation is not a risk because it does not look like a deleterious event that may happen, say a 25 per cent chance of a wall collapsing. It is, however. Along with the 25 per cent case is a 75 per cent chance that the wall will not collapse, and so in terms of the unresolved option primitive, and substituting 'demolish' for 'collapse' we have an unresolved option of to demolish, or to not. Further, the primitive also provides a way to model a set of options: demolish, repair, replace or leave. What matters is that the set of options is complete so that the probability of something happening, even if it is nothing, is 100 per cent.

Further, some may say this is not a risk because the options will be resolved in due course. One would hope so, but I have always found some even after a project has been given the go-ahead. As a result, because the estimate will be based on an assumption of the choice having been made, I include the others in the risk assessment because in a funder-centric view their continued existence creates possible reasons for future overspending.

Primitive 4 Emerging Cost Chain

There are many things on a project that are just risky: they are palpably capable of going wrong, and if we can foresee such situations then we need appropriate primitives so they can be put into the risk model and influence the cost function.

This type of risk is most common where the works of a project involve the possibility of the discovery of something that might thwart the project, for example that some cables that need to be diverted are too frail, or that a batch of samples are contaminated. Often a facility, service or source of material intended for use by a project has been taken by another project, or planning permission is given for a new building to go up on a site earmarked for use as a temporary construction site, or another project has bought the entire product output of a manufacturer for a period of time.

The primitive needed for these is the emerging cost chain, which because it is less immediately obvious than the previous three primitives, I will explain using an example based my own experience.

A high-speed railway train operates between A and B to a strict timetable. Failure to keep to the timetable results in the railway company being fined. Railways inevitably wear out, so if the bearings on a bridge have done just that, the train may experience a heavy jolt as it passes over it. This jolt is so severe that the company is concerned it will cause a derailment and so decides on safety grounds to impose a speed restriction over the bridge. However, the stretch of speed-limited running is sufficient for trains to be persistently late and so the company has to pay penalties. A decision is made to replace the bridge, which will cost less than years of fines.

However, it proves difficult to specify the works precisely. Certainly the section of the railway over the bridge will have to be temporarily removed and then the deck exposed. The deck-retaining pins will have to be cut so that it can be lifted out in one piece. The new deck will have to be lifted in. It has been noticed that the drainage outfalls around the bridge are dry and so it is thought that the drains may be blocked. If they are then the ground behind the abutments of the bridge may be waterlogged, the pressure from the weight of the water being sufficient enough to spring the abutments inwards slightly when the old deck is removed (Figure 2.5).

If this happens then the new deck will not fit until the abutments are then disassembled, the waterlogged material in the embankment dug out and replaced, the drains are repaired and the abutments reassembled so that the gap between is wide enough for the new deck to be lifted in. The risk looks like that shown in Figure 2.6.

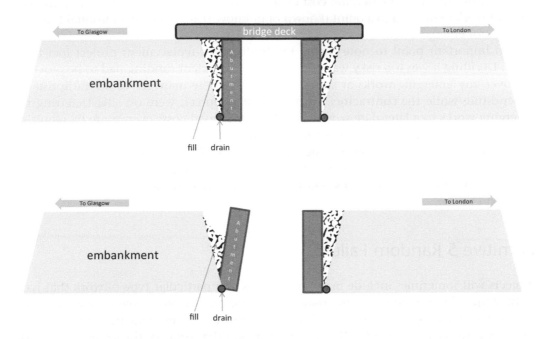

Figure 2.5   The bridge deck replacement works, before and after the removal of the deck

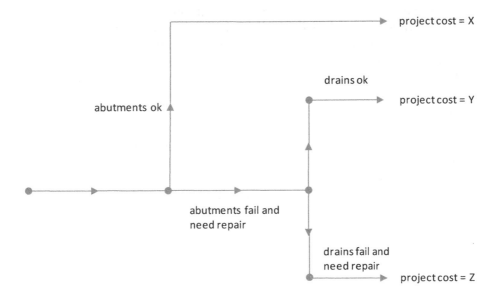

Figure 2.6   Emerging cost chain for the bridge replacement works

Reading from right to left: the old deck is removed. Then either the abutments will spring or they will not. If they do not, the new deck is fitted. If they do, the abutments are taken down, the embankment and drainage repaired, then the abutments are rebuilt and the new deck fitted.

The primitive for an emerging cost chain is thus a modelling element that allows a piece of work to proceed to a point (known or unknown) at which an additional task may or may not emerge.

An important point to note is that the bridge deck replacement project had to be done. Deciding it was too risky was not an option. Sufficient funding had to be provided to cover any emerging works at short notice so that the manager could authorise the expenditure while the contractors, with all their resources, were on site. Deferring the emerging works to a later date would only have increased costs even more by requiring everyone to return and set up the work site again.

Emerging cost chain primitives are often doubles; the emerging work requires both additional funding and additional time, and that additional time can translate into further additional costs. We will see how this is dealt with in the risk model example in the next chapter.

Primitive 5 Random Failures

Projects will sometimes include many repetitions of a particular type of work that have to be done within a repeated time frame: 20 IT stations set up per week; railway repairs completed at 5 sites over a weekend; 100 just-in-time deliveries made per day, and so on. There is always a risk that all will not go to plan and that some of the set-ups, site works and deliveries will not happen.

We therefore need to have a primitive that allows for a 1 in X failure in cases where the failure will incur additional costs.

The primitive must have a subtle but important behaviour. Say 1000 deliveries are made per week, of which 1 in a 100 is the probability of a failure to deliver just in time. I am not saying exactly 100 of the 1000 will fail, just that these failed 100 need to be funded in some way, say by over-purchasing 100 more (of which ten will then fail to be delivered just in time, and so on). The failure rate of 1 in a 100 is the historical average failure rate of all deliveries made to date, and in any one week the deliveries may go very well and none fail at all, or on the other hand, they may go extremely badly and more than 100 of them may be late.

The primitive has therefore to model risk in a way that allows the number of failures to vary in the short term but which maintains a long-term average. Figure 2.7 shows the number of failures per week of a repetitive process, such as the just-in-time delivery of component parts. The number per week varies between a minimum of 4 and a maximum of 17, but overall, the long-term average is 10.

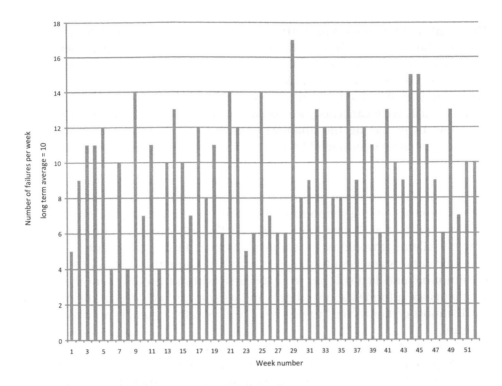

Figure 2.7   Random failures of repetitive processes

Primitive 6 Not Modelled

Before moving onto an important subset of primitives used for the modelling of time, I would like to incorporate one that has no mathematical modelling associated with it all, the 'not modelled primitive'.

There are three reasons why a risk cannot be modelled:

1.   When to model it would be pointless, which is not to say the risk is not a risk.
2.   When we do not know enough about it.
3.   When we understand the risk but do not yet know how to model it.

At the time of writing this I am an analyst on a major infrastructure project that in the prevailing economic climate of increasing government deficits and collapsing banks is at serious risk of being cancelled, or since that may be too politically embarrassing, of being deferred while other ways to design and build it are considered – anything that would postpone the start of construction with its implicit major expenditures would, I am sure, be attractive to the funders. Further research and option evaluation would be much more affordable to them than starting detailed design and construction.

However, for the company that is promoting the scheme, cancellation or deferment represents real risks to its continuance as a business, and as a risk it expends a considerable amount of time and effort by working with the Government and its agencies to convince them of the value of the project.

The point that I wish to make is that these risks, cancellation and deferment cannot be modelled but are crucially important for the business to manage. They cannot be modelled because we cannot judge the probability of cancellation nor the time frame in which it may occur. It may be sudden, and strenuously resisted by everyone involved until that moment. It would represent a cusp in the continuum of the project's existence. Furthermore, if it were cancelled, we could not be confident about what conventions should apply to predicting the financial consequences. Should it be just the cost of terminating my client's contracts? Should it include a cease-trading scenario or a go off and do something else scenario? Should it include the value of the opportunity lost? Here we have risk that needs considerable management attention but which we cannot model.

Say we only became aware of these risks – cancellation and deferment – during our research within the funding community about the risks to securing the long-term finance the project requires, and that we found ourselves the only people to know of this possible threat to the client? It would be a clear dereliction of duty to the client to not mention it in the calculations and resulting reports simply because they could not be modelled. What cannot be modelled still has to be declared within the outputs of the risk process so that clients are informed about what is and what is not included in the cost function. They will almost certainly need to manage it successfully – there is no funding to cover its exposure.

A risk model – the assembly of all of the risks whose behaviour is represented by sub-assembly of primitives – has to be built on a presumption that given the right amount of time and money the project can be completed, and that any risk to the contrary, in which the unwanted outcome is that the project will not be completed, is excluded from the calculation. Crucially, such exclusions must be made clear when reporting the analysis results.

There are other reasons for not modelling risks: perhaps how to model them is not known, or their modelling is still being researched. You may come across a risk behaviour that is not covered by the approach presented in this book and so the risk in question is not modellable pending completion of your own analysis of its behaviour. These must all be declared when then analysis is reported.

No matter what the risk is, whether it is modelled or not, the funders need to know about it.

Primitive 7 Time-dependent Costs

Even if the analysis of the project shows there are no uncertainties in quantity and rates, no amended options, no emerging cost chain and no random failures, in my experience there will always be at least one risk of the time dependent cost type. I have also found that it will always be in the top 10 largest risks of a project. Time dependent costs are always present, even if it is only the cost of the project manager (though the category usually embraces rental of space and machines, hire of labour, interest charges, penalties for lateness and the marching army of project services). These will all continue to fill in timesheets, booking time to a project even if its planned completion date has passed.

If a project overruns then it will almost always cost more. I know of no project that has not overrun that has not also been overspent, though it is possible for a project to overspend by paying for acceleration of the work, by hiring of additional resources and by offering incentives and bonuses in order not to overrun.

To assess the effect of time-dependent costs we first need to have the calculation project planners use to work out how long it will take to complete a project – one that is equivalent to an estimator's calculation of how much a project will cost. This is more complicated because whereas overall costs are a columnar summation of the products (the quantities and process of each item to be used), the timescale required for a project is not the summation of how long it takes to complete each individual task. Rather it is that summation less the time required for every activity that is planned to be carried out concurrently with another, subject to the proviso that the removed items cannot then be formed into a chain of activities whose end to end time is longer than that of the remaining set. If they do, this extracted sequence becomes the one that determines the timescale of the project. It is easier to see with an illustration (Figure 2.8).

In Figure 2.8, the planner has identified a network of 12 tasks for the construction of a production facility. Each has both a duration and a place in the logical order in which the tasks are to be carried out. Many can be done concurrently as shown by the parallel paths in network. One path is red – this is the longest path through the network in terms of overall duration and so determines the duration of the project.

A network like this forms the basis of the time-dependent cost calculations. To predict the effect of them we now need a set of time-related primitives.

Primitive 7.1 Duration Spreads

This is the most commonly used time risk modelling primitive.

The timescale calculation uses fixed durations for each component task. These are sometimes known as spot durations, and the duration spread is obviously a range on these, for example:

- 6 weeks plus or minus 1 week, i.e. 5 to 7 weeks
- 12 months plus or minus 25 per cen,t i.e. 9 to 15 months

The ranges need not be equal about the spot, for example:

- 6 weeks, plus 3 weeks minus 1 week, i.e. 5 to 9 weeks
- 12 months plus 25 per cent minus 50 per cent, i.e. 6 months to 1 year.

The spread is continuous in these examples, in that the actual duration could be found to have been anywhere within the range once the task has been completed. The spreads also include the opportunity for the eventual duration to be shorter, as well as the risk for it to be longer.

The spread duration primitive is initially suited for modelling situations in which production rates may vary due to laziness or enthusiasm; rain or sunshine, hard or soft ground, ease or difficulty of access, quickness or slowness of response times. It is also useful where some of the details of a project plan are not yet known, an example being for a plan that is mapped out in weeks, how many working days there will be in a working week: five, six or seven?

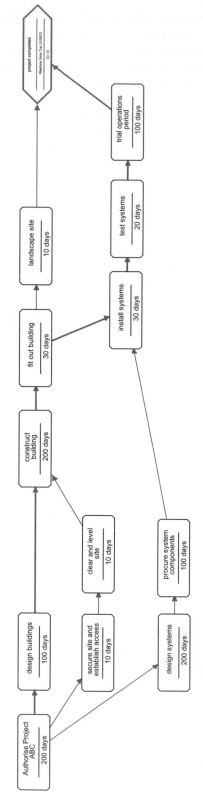

Figure 2.8 An example of a logic linked plan

Digression: Why Time Risk Assessment is Important

Here is an interesting calculation intended to illustrate a point for modelling time risks on the grounds that most cost risk analyses will be wrong if the model does not include time risks.

Say we have a project plan that has only two tasks in it, and each of these is planned to last four weeks. One task is to build a house and the other to layout the gardens. The plan would look like this (Figure 2.9).

Since the tasks are concurrent, the overall duration of the project is four weeks. Both tasks must be finished before the owner will accept the property. Suppose a risk analyst is asked to assess the project, and decides to ask the builders if they can foresee any reasons why their task may take longer than planned. They reply that if the project is commissioned in the summer then the dry weather, long days and desire of their staff to top-up tans means the house could be built in two weeks. If on the other hand it is commissioned in winter, it will take six weeks.

The analysts then put the same question to the gardening contractors. They say that if their new digger is in the current batch being assembled at the factory then early delivery is likely, which in turn means they could be finished in two weeks. But if it is in the next batch there will be a month's delay and so their work will be completed in six weeks.

The project plan now looks like Figure 2.10.

Figure 2.9   A simple plan

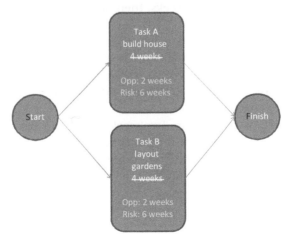

Figure 2.10 A simple plan with risk and opportunity spreads on duration

A supplementary question to ask each contractor is that if the timescales are actually two weeks or six weeks, why then did they offer four? A simple answer could be that the project planning tools they use only allow a single-digit number as the duration and not a set of possible values, the choice of which would depend on conditions prevailing at the time.

Another response could be for them to point out the commercial realties of the situation. To base their tenders entirely on the worst case, maximum task durations of six weeks, would exceed the client's requirement so they would more probably lose the work to someone else.

Equally it would be foolhardy to tender a delivery date based on all the minimum durations – two weeks. A balance has to be struck, one that in the judgment of the bidding teams offers a good chance of delivery in a timescale that suits the buyer and that also offers a good chance of an acceptable profit. Here the commercial managers of each contractor decided four weeks was that balance point.

The analysts must answer this question given the reality is two or six weeks and not four, when does the project finish? I am about to play fast and loose with the numbers here in order to illustrate a point.

These are four possible outcomes. The assessment is shown in Table 2.2.

Table 2.2   Possible outcomes for the simple plan with risk and opportunity spreads on duration

Task A build house	Task B layout gardens	Overall
2 weeks	2 weeks	2 weeks
2 weeks	6 weeks	6 weeks
6 weeks	2 weeks	6 weeks
6 weeks	6 weeks	6 weeks

If both A and B were to take two weeks, then because they are concurrent tasks, the project will also finish in two weeks.

If however, B takes six while B takes only two, the project will finish in six weeks because both A and B must be completed first. The same is true if A takes two weeks while B takes six: the project will be finished in six weeks.

Finally, the worst case is when both A and B take six weeks. The overall completion is also six weeks.

There are only four possible results. Two weeks can occur once as the outcome, but six weeks can occur three times, thus it is three times as likely to occur. If this project was to be repeated many times, i.e. a number of houses were to be built, the average time to complete each of them would be as follows.

Average time to build a house:

= (1 in 4 chance of 2 weeks) + (3 in 4 chance of 6 weeks) = (1/4 × 2) + (3/4 × 6)

= 5 weeks

Not four weeks. The presence of risk on a task (that each task may take six weeks) can increase the duration of a project. The presence of opportunity on task however (that each task possibly may take two weeks), does not necessarily mitigate the risk: the project will still take a week longer. There is a ratchet-like effect in time risk assessment that can make it difficult to recover time lost to risks by realising opportunities to save it elsewhere. It is different for cost risks in that overspends can and usually are recovered somehow, even if it is only the painful experiences of refinancing and claims settlement. But time is different because it is a currency granted by another stakeholder, the universe if you like, and once it is lost it may not be recoverable by anyone involved in the project.

It is never the case that with time risk, some you win and some you lose, or that some things finish early and some things finish late. The losses will prevail, and the time risk exposure will accumulate steadily and relentlessly. Through its leverage on time dependent costs, it will have a subsequent impact on total project costs. I have never known a properly evaluated time risk to be outside of a project's top ten risks.

You may argue that since the result can only be two or six weeks, what is the point of knowing that the average is five weeks, a duration which is not among the possible outcomes? Though the example was a disingenuous intended to make the point that time risk ratchets up – knowing the average duration *can* be a useful performance indicator in practice. For example, say the time a contractor takes to complete a common piece of work begins to diverge from the average. Clearly there is something wrong. Perhaps the equipment being used is worn out, or absenteeism is rising. The trend away from the average (which may never in itself be a possible, actual, duration of the job) can be analysed to assess the financial consequences, and used as the basis for investment or management action.

Primitive 7.2 The Fuzzy Overlap

Many types of work can be overlapped. For example cutting the grass and rolling the grass; washing the car and drying the car; plastering the wall and painting it, can all be done be done at the same time provided drying follows washing and so on. There are two types of overlap, the lead and the lag. Plastering can be considered as leading the painting or painting lagging the plastering. Either way there is a temporal offset between the start of one and the start of the other.

In the planning world, in which tasks of fixed duration are to be done in a fixed sequence, any overlaps, like durations, will also be of fixed durations. The question a risk analyst should ask is how long does it take plaster to dry before it is suitable for painting? The answer is that it all depends on how wet it is and how dry the ambient conditions are. It could be anywhere between one and five days. Thus a fixed lag of, say, five days between painting and plastering in the planning domain (Figure 2.11) would become a lag of between 5 and 10 days in the risk domain, a fuzzy overlap (Figure 2.12).

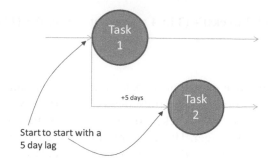

Figure 2.11 A lag between two tasks

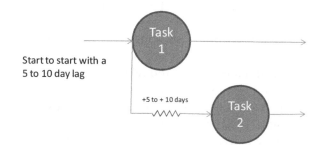

Figure 2.12 A fuzzy overlap between two tasks

There are many real-world situations where a fuzzy overlap would be the right form of risk primitive to use to model variable amounts of time such as:

- Waiting for people to turn up before proceeding
- Waiting for machines to become available before starting
- Waiting for software testing to finish before resuming
- Waiting for material to set before continuing.

Any of these could cause an extension of time that could ripple through the logic of the plan and drive out the end date of the project.

Digression: Why it is Useful to Keep Risk Primitives Distinct From Each Other

There is a close similarity between the spread duration and the fuzzy overlap in that both describe a period of time whose length is uncertain. In the former, time is given over to work, and in the latter to waiting. The similarity does not however mean they ought to be coalesced into a single primitive called, say, variable period of time, because the hidden purpose of primitives is to help us identify the risks during discussions with engineers and managers, and they may well will distinguish between non-working and working time. An awareness on their part that a risk model can accommodate their logical distinction may stimulate their exposition to us of what the risks are. The analyst may then model the spread durations and fuzzy overlaps using the same mathematics, but that need not be a concern of their interviewees (if they so wish it).

This separation of the logical description of a primitive from the physical nature of its mathematics – its defining equations – is a valuable aid to ensuring that those who contributed their knowledge about the risks to the modelling process accept its results. What they said, in the way in which they said it, can be seen in the model.

When interviewing project staff working on the obtaining of planning permissions about the risks in their work, the concept of rework cycle connects to their experience in a way that talk of interruption would not. For project staff engaged in construction, it is the opposite way round – progress is interrupted and possible occurrences and reasons are readily foreseen. But rework? That would almost be tantamount to an admission of not being able to get it right first time, a possible offence in the Courts of Project Management.

Furthermore, using logical primitives in the model is consistent with one of the most important tenets of good analytical work, which is that the model should fit the problem and not the problem the model. Using simple risk models when sophisticated ones are necessary is poor practice, especially when investment decisions are to be made on their results.

An analyst who asks a team of construction workers if they know of any variable time periods that may arise in what they are about to build may well impress the audience and establish an intellectual authority, but it will be a self-regarding one that is more likely to dampen the conversation than to stoke it up. All an analyst has to do is to get the right answer, and asking if anyone knows reasons why the task durations may be more or less, or why leads and lags may not be as shown in the plans is an approach more likely to find them than one which presents the esoteric of risk modelling to an audience who are not actually there to buy it. Not only should the analysis fit the problem, so too should the analyst.

Primitive 7.3 The Emerging Task and the Disappearing Task

Every logic lined plan for a project has three limitations. Each activity must be:

1. Carried out only once
2. No longer than the duration specified
3. Executed in the sequence set out in the logic of the plan.

The second point is not always true. Activities that are not on the critical path of a project, the one that determines the overall project duration, can accommodate a lengthening of their duration so long as this additional time does not change the path through the project that determines the overall project duration to be the one with the newly extended task.

Thus in Figure 2.13, the Build car park activity has a float of six weeks that can be used to absorb delays in its construction. The path that determines the overall duration of the project will remain the one in red that passes through the activities Build IT centre and Install IT.

Suppose however that the task of building the car park was subject to a risk that upon completion of its design, there is a risk some of the land required will be discovered to be just outside the boundaries of the land acquired for use under a compulsory purchase order. If a price for the land cannot be agreed with its current owner then an addition to that purchase order would have to be sought, which could take up to a year to obtain. If this situation transpired, the logical path through the plan with the task of building the car park would determine the duration of the project.

What if the planner had mistakenly assumed that all the land needed for the project was purchasable under a compulsory purchase order? Realisation of the possible need to acquire more could be an emerging task. Indeed it could even be two mutually exclusive ones – either a deal with the owner is struck, or an application has to be made for an extension to the compulsory purchase order. The situation is shown in the following figure (Figure 2.14).

The emerging task primitive allows the risk analyst to incorporate into the risk model things that may or may not have to be done, unlike the plan in which they all have to be.

Of course it may be the case that the planner has been cautious and has put into the plans all tasks that *may* need to be done as well as those that must. If those that may be done are not on the critical path then including them has no effect on the project duration, they do not result in misleading advice about the completion date. But plans are often speculative and driven by an end date imposed by a client. My experience is that plans, therefore, tend to contain fewer tasks than actually need to be done, showing instead just enough to demonstrate the end date can be achieved without any glaring omissions from the tasks and illogicality in their sequencing. This is surprisingly common. A construction sequence in a plan may be perfectly logical and accurately timed, but the preceding procurement phase will be 'High Level' and its timescales optimistic.

Though it may cause many planners to take umbrage, only occasionally have I seen plans in which the planner chose the duration of tasks based on evidence of productivity rates, the size of the job and the size and skill of the resource pool. Often task durations are compressed to fit into an overheated programme of hopefully seamless interfaces and unhindered progress in order to concur with what the client says must be the completion date. It seems to me that whereas cost estimators are allowed to work out the cost, planners never are the timescales but instead are told them, and in that observation, if you are a planner and are irritated to read these words, I hope you will sense some genuine sympathy.

As before, the purpose of the primitive is not to enforce a mathematical distinction between types of risks but to help identify them within the real-world experience of project people. And as before, the next chapter will illustrate the modelling with some examples.

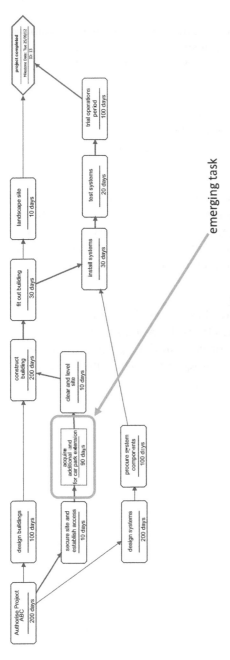

emerging task

Figure 2.13 The plan with an emerging task that alters the duration of the project

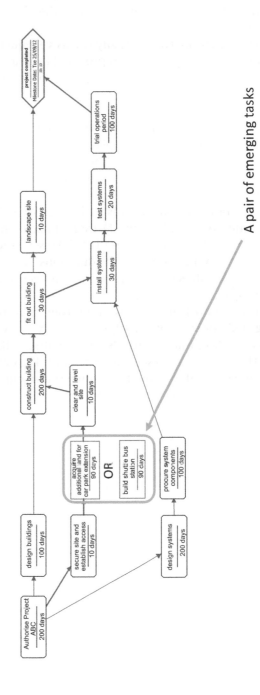

A pair of emerging tasks

Figure 2.14 The plan with a pair of mutually exclusive emerging tasks that alter the duration of the project

Primitive 7.4 The Emerging Logic Primitive

In the previous section, the emerging task appeared on an existing path through the plan. However, there is another case, the emerging path comprising a sequence of tasks (Figure 2.15).

Here there are two mutually exclusive sequences of work. A system is either developed in-house or it is specified and procured from an external supplier. Only one will be done. If the plan had assumed it was going to the in-house development, then if a decision was subsequently made to procure the system from an external supplier, this sequence of work would be an emerging logic and the intended logic would be abandoned.

The solution to modelling the emerging logic scenario is to recognise that in risk modelling, unlike in project planning, both sequences can be present in the calculations. If the sequence is not taken, either the in-house development or the external procurement will have no effect on the overall project timescale, and so the duration of tasks in it can be set to zero.

And how can one do this with dice? We can use a two-sided one, a coin. Heads and the durations of A, B, C and D are set to zero, implying external procurement will be the decision; and tails, E, F, G and H will be set to zero, implying in-house development will be chosen. Whichever it is, the durations of the tasks on the other sequence, which may possibly be subject to other risks themselves, will help determine the overall duration of the project.

If the path through A is more likely than the path through E then the coin can be weighted, for example by stipulating only every second head switches on the E sequence and the odd-numbered heads and the other tails switch on the A sequence.

This primitive is especially useful for IT systems projects and for projects that have a research content, because these often spawn emerging programme logic if new discoveries are made while they are in progress.

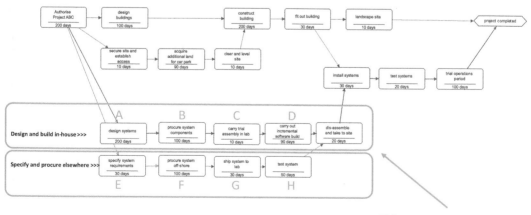

Either A-B-C-D or E-F-G-H

Figure 2.15 A plan with an emerging logic path

Primitive 7.5 The Rework Cycle

Of all of the risk modelling primitives I used in risk analysis, this is the one that most strikes a chord with project people because they are convinced they could complete the project on time and to budget if only the client could let them get started. In a project plan each task is intended to be done once and only once, but the rework cycle primitive allows the risk that a task will have to be repeated (wholly or in part) to be included in the risk model.

The interface between project funder and project doer is a thicket of nettles, brambles, cowpats and unexploded ordnance. This is always the case where the funding is by the public sector, so much so that I wonder why neither side properly allows for it in their plans. Instead, the lobbing of the ball of submission for permission, and its subsequent return, rejected and deflated for reworking, is the enervating lot of the project team.

The business of gaining approvals is riven with rework cycles. So too is the work of quality control with its not altogether uncommon rejection of materials and completed work, and also the integration of systems with its timings, tweakings and re-tests. Figuratively a rework cycle looks like Figure 2.16.

Like the other primitives, it adds a quantity of time, and possibly a spread quantity to the duration of the underlying task. Like the emerging task, there is a probability of it happening. There is also a probability of it happening again (Figure 2.17).

This is the multiple rework cycle, which is characterised by a reducing probability of the second, third, fourth … cycle occurring.

A rework cycle can be a full or part repeat of its parent. This is the usual case with submissions of paperwork seeking permissions, because generally clarifications and corrections do not require recreation from scratch, only revision. Having said this, it is important to be aware when modelling that the length of the rework cycle is often not set by the time required to make the revisions, but rather by the frequency with which the adjudicators meet to consider submissions. But as anyone who has ever missed a flight can tell you, the next free spot on their agenda may never materialise.

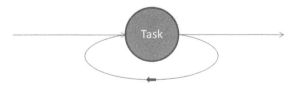

Figure 2.16 A rework cycle

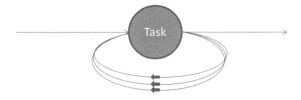

Figure 2.17 A multiple rework cycle

Primitive 7.6 The Interruption

Many activities in project plans have durations that presume unhindered progress. However, nature can be inconsiderate and the context in which the task is being done is often not stable and benign. Work sites are often at risk of being closed due to inclement weather, as is progress to being stopped by power outages, labour strikes, holidays and legal injunctions.

In such circumstances the modelling of the task duration needs to allow for additional time before work can resume (Figure 2.18).

As with the other time-modelling primitives, an element of additional time needs to be introduced together with a probability of it being incurred. Though here the time is not added to the end of the task but at some random point within it; it could be today, but could also be tomorrow or next week.

Note that the probability of an interruption may change over the planned time frame of the task, for example a project to build a sea wall over a year is more likely to be interrupted when tides are high or storms prevalent.

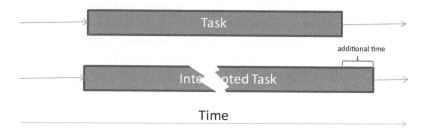

Figure 2.18 An interrupted task

Primitive 7.7 The Disruption

This primitive is used to model the variability in duration of activity caused by the risk of an interruption that, while it does not suspend work, hampers its progress (Figure 2.19).

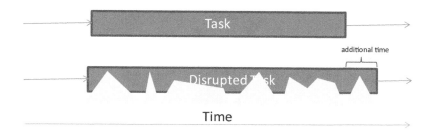

Figure 2.19 A disrupted task

Good examples of impositions placed on projects after they have commenced are:

- Restrictions on working hours
- Noise and dust levels
- The sudden curtailment of delivery and disposal routes.

Primitive 7.8 The Productivity Growth and the Days-to-Complete

Often the activities in a plan comprise several repetitions of a standardised task, for example, setting up a workstation in a call centre, or fitting a section of pipe in a project to build a pipeline. Standardised is probably too rigorous a word. Familiar would be better, because it would imply there may be a degree of variation in the time to do the job caused by changes in staff circumstances or the workplace environment.

Suppose there is, as there might well be if the job is repetitive one, some historical data about how many of these tasks can be done in a day. For the purposes of illustration let the number of workstations that can be installed in a day be three, four or five according to circumstances. At the end of day one we would expect to find what we see in Figure 2.21.

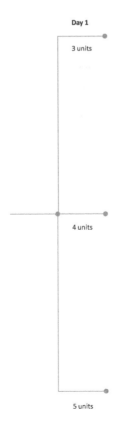

Figure 2.20 Possible production on day 1

Day 2's production could likewise be (Figure 2.21).

However, Day 2 follows after Day 1, during which three, four or five stations may already have been set up. Day 2's production follows on from Day 1, whose actual production we do not yet know because as in all risk analysis, we are assessing situations that have not yet happened. There are three possibilities (Figure 2.22).

Day 3's production follows that of Day 2, whose actual production we also do not yet know (Figure 2.23).

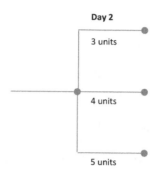

Figure 2.21 Possible production on day 2

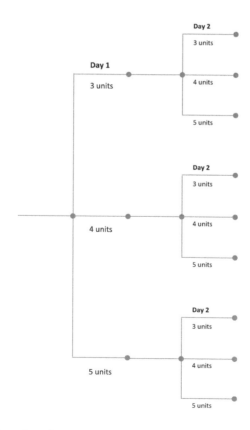

Figure 2.22 Production for day 1 and day 2

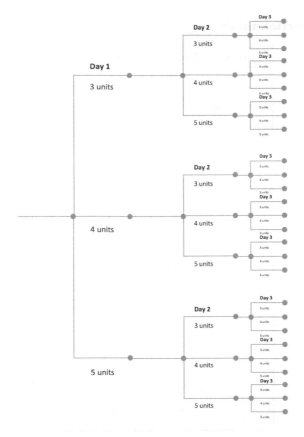

Figure 2.23 Possible production for days 1, 2 and 3

The inclusion of Day 4 would fill this page with possibilities, so let me constrain the analysis to three days. By adding the number of installations along each branch, we can see the totals that may be achieved after three days (Table 2.3).

The worst possible outcome is three days of three giving a total of nine overall, and the best is three days of five, giving 15 overall – that is assuming zero units is never the total installed in a day, which I suppose is not the best assumption to make, but then 0 or three or four or five would expand the figure off the page. A histogram of the number of occurrences versus days would be as follows (Figure 2.24).

Table 2.3   Production totals after three days

Day 1	Day 2	Day 3	Total
1	1	1	3
1	1	2	4
1	1	3	5
1	2	1	4
1	2	2	5
1	2	3	6
1	3	1	5
1	3	2	6
1	3	3	7
2	1	1	4
2	1	2	5
2	1	3	6
2	2	1	5
2	2	2	6
2	2	3	7
2	3	1	6
2	3	2	7
2	3	3	8
3	1	1	5
3	1	2	6
3	1	3	7
3	2	1	6
3	2	2	7
3	2	3	8
3	3	1	7
3	3	2	8
3	3	3	9

This is a spread of 9 to 15 units, though not a uniform one. Instead it is biased towards 12 units being the most likely outcome after 3 days work.[1] This too is a dice, albeit a seven-sided one with faces showing 9, 10, 11, 12, 13, 14 and 15 dots. It is also a loaded dice that is biased towards 12 units of production occurring most times, with the other production totals appearing less often, in accordance with their reducing probability of occurrence.

---

1   This analytical method is called Semi-Markov modelling. For a full description see http://en.wikipedia.org/wiki/Markov_chain, http://en.wikipedia.org/wiki/Semi-Markov_process and linked references.

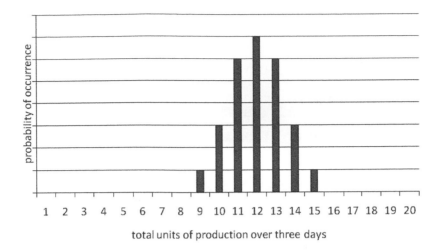

Figure 2.24 Distribution of production totals after three days

Having such a primitive may be useful in two ways; for modelling learning curves, and for modelling the number days needed to complete a given number of repetitive tasks.

Taking the learning curve first: suppose we have an expectation that as the installation team becomes increasingly familiar with the computer room (unwrapping the packaging, the cable ducting) the number of installations per day would increase from, say, 3-4-5 to 4-5-6 to 5-6-7, where it would settle. This gives the outcome shown in Figure 2.25.

Second, the Primitive can be used to model the number of days needed to complete a quantity of repetitive work (Figure 2.26). This can be achieved by adding additional days until a target number of installations, completions or similar, say 20, appears as a possibility (Figure 2.27).

Notice how, though this evaluation shows 20 completions are possible, it is not certain and it could well be fewer: 17, 18 or 19. More days could then be added to the model until 20 is 100 per cent certain (Figure 2.28).

Twenty-two is now the minimum outcome and so completing twenty installations is therefore assured.

The primitive allows the analyst to derive a minimum to maximum number of days for the completed task, together with a 'most likely' number and the probabilities of each happening.

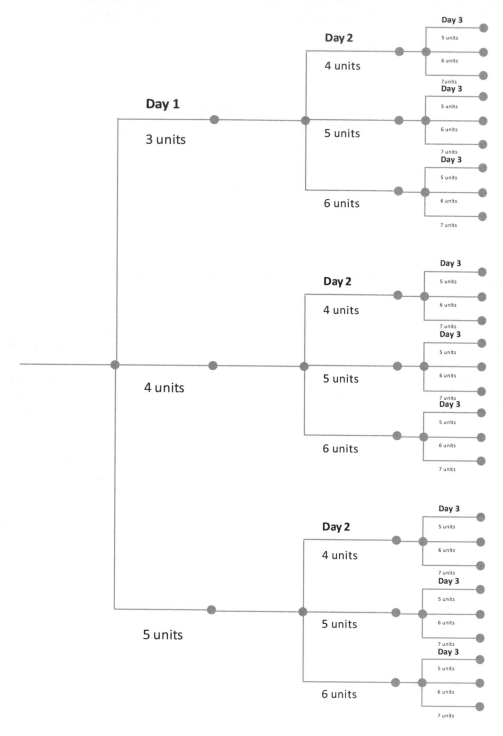

Figure 2.25 Modelling a learning curve

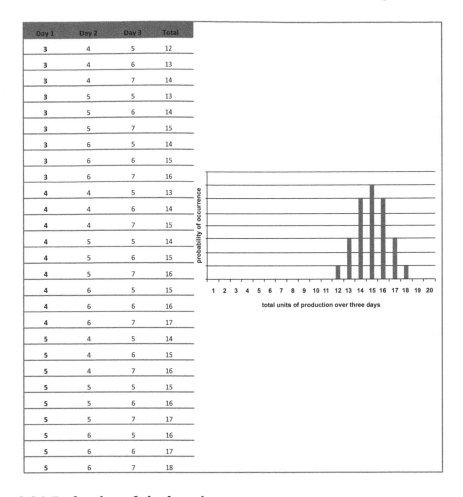

Day 1	Day 2	Day 3	Total
3	4	5	12
3	4	6	13
3	4	7	14
3	5	5	13
3	5	6	14
3	5	7	15
3	6	5	14
3	6	6	15
3	6	7	16
4	4	5	13
4	4	6	14
4	4	7	15
4	5	5	14
4	5	6	15
4	5	7	16
4	6	5	15
4	6	6	16
4	6	7	17
5	4	5	14
5	4	6	15
5	4	7	16
5	5	5	15
5	5	6	16
5	5	7	17
5	6	5	16
5	6	6	17
5	6	7	18

Figure 2.26 Evaluation of the learning curve

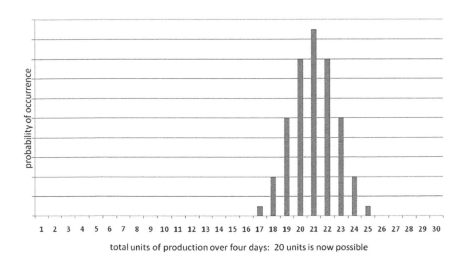

Figure 2.27 Adding days until the target becomes a possibility

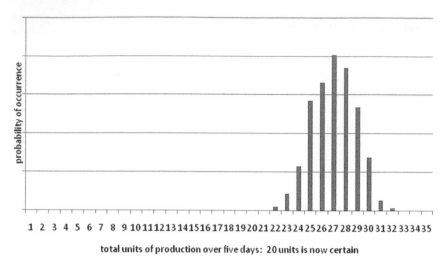

total units of production over five days: 20 units is now certain

Figure 2.28 Adding days until the target becomes a certainty

Summing Up

We have been through the 14 primitives I use in my approach to calculating the funding a project needs. They are:

1. Primitive 1: Spread on quantities
2. Primitive 2: Spread on price
3. Primitive 3: Unresolved option
4. Primitive 4: Emerging cost chain
5. Primitive 5: Random failures
6. Primitive 6: Not modelled

Time-dependent costs

1. Primitive 7: Duration spreads
2. Primitive 8: Fuzzy overlap
3. Primitive 9: Emerging task and the disappearing task
4. Primitive 10: Emerging logic
5. Primitive 11: Rework cycle
6. Primitive 12: Interruption
7. Primitive 13: Disruption
8. Primitive 14: Productivity growth.

Primitives 1 to 7 are for cost risks, and 7.1 onwards for time risk; 7 is a cost primitive, to model this requires the subset of 7 × time risk primitives.

In bad analysis, a prescribed method is used to analyse a problem no matter what its nature. You may rightly think this preposterous and yet it is common. It beggars (my) belief that the problem of how much funding a project will need is still most often answered by summing the products of the quantities, rates of the component labour and materials and adding on mark-ups, and further, that the amount of time needed to complete it is calculated using a logic linked plan, if it the duration not already been imposed by the client. These are prescribed methods that do not fit the funding problem of finding out reliably how much money and time will be needed.

My argument is that the funding has to cover the costs and the risks. However, too often the solution to this is to make the problem worse by carrying a poor risk assessment, one that uses only a single form of primitive for every risk, no matter what its behaviour: the triple-point estimate (Figure 2.29).

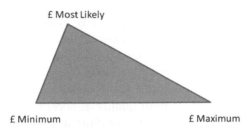

Figure 2.29 The triple-point estimate

Like costing and planning, risk often makes the problem fit the available analysis.

The advantage of having the broad set of primitives described so far is that they allow the analysis to fit the problem, both mathematically and conversationally. I hope you can see this although I have yet to show how primitives are modelled in the spreadsheet world, even though many instances of differing primitives do end up as being modelled similarly.

The key point is that by giving the analysis the flexibility to match the experience of the engineer and the manager to the expertise of the risk analyst, the business of eliciting good risk information is made far easier, as is the acceptance of its results. Imagine an experienced project manager explaining that there is a risk that 10 20 per cent of the planned weekend workings will be lost to labour absenteeism, resulting in anything between a half and a total loss of productivity depending on whether the staff are critical or not, and then asking what is the minimum, most likely and maximum cost of that and the probability of it happening? The prescriptive approach does not fit the explanation given, but a random failure primitive does.

The value of using primitives that make experience and modelling coherent with each other can be seen in another way. The power of a rich set of primitives is not that they allow the analyst to mould the analysis to the experience so much as they allow the experience to tell the analyst about the problem in a way that the analyst can use.

Turning the Primitives into Numbers

The purpose of risk modelling is to show the potential costs of a set of risks. This is something I call the risk exposure. Strictly speaking this ought to be 'risk and opportunity' modelling, but it is clumsy phrase and I prefer the conciseness of 'risk'. In what follows, where I write 'risk' I do mean 'risk and opportunity', and thus at a stroke I confound those who think risk work is flawed because it concentrates on the negative and does not see the positive. As old men know, opportunity is like snow in spring whereas risk is a block of concrete all year round. It is funny how you only ever see young 'uns taking up the role of opportunity manager.

The reason for wanting to know the risk exposure is to add it to the cost estimate to find the cost function, and from this to ascertain the amount of funding needed to complete a project on time and to budget with a desired level of confidence.

As the annotation on the right-hand side of Figure 2.30 shows, the cost of project A at a 90 per cent level of confidence that this will not be exceeded, is £17.3M.

The condition, 'that it will not be exceeded', is important because the probability that it will be *exactly* X amounts of pounds is low, a percentage so low that if announced publicly, can cause a lot of confusion amongst funders. It is as low as the probability that the project will cost exactly £X + £1 or £X + £2 and so on.

However, the probability that project A will cost no more than £X + £2 is three times as high as £X because £X + 1 and £X both qualify as acceptable outcomes. I am ignoring values less then £X here simply for economy of illustration.

The funding community does not mind, indeed is pleased, if a project costs less, as implied by the description 'will not exceed £X'. What really worries it is if the project looks like it is going to cost more because more may be very difficult, if not impossible, to find. The emergence of such a situation can be extremely disruptive to progress. In my day job, if the total expenditure on the project is forecast at any time to rise above the level I have predicted should occur at a lowly 50 per cent level of confidence (which is quite a low value in the range of possible costs), then no matter whereabouts in the project plan they are, staff must interrupt their daily grind of designing and constructing things to the devising, presenting, negotiating and implementing of a recovery plan to get the costs back down to the 50 per cent level. Moreover, no extra time will be given to the project to compensate for that loss while tending to the interruption because that could put up costs through the effect of time-dependent charges.

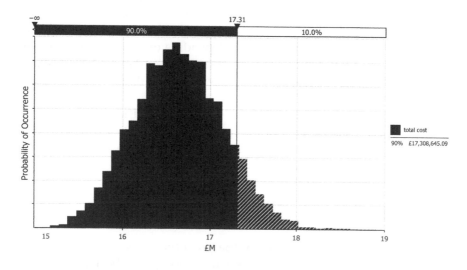

Figure 2.30 Cost function for the project in the example model

The potential for a project to cost less than £X allows the possibility of introducing management incentives in the form of bonuses. A target-cost less than £X may be introduced with any savings on cost being shared between the funders and the project management team. Caution is needed, however. This strategy can give rise to the phenomenon of risk analysts unethically boosting their calculations to create an artificially inflated exposure level that is then funded, the £X, but which because exaggerated, is never likely to occur in reality. When the final accounts are drawn up, the expenditure will be all the more likely to be less than that funded, yielding unearned bonuses. The only ways to discover and stop this practice are either to have the risk assessments independently scrutinised, or not to have the risk assessments carried out by the project promoters, but instead by the project funders themselves.

To compute the range of costs for a project we have to have a risk-modelling tool. The job of the tool is to provide the electronic dice and simulation operations needed to compute the risk exposure that can in turn be added to the estimate to give the cost function. The electronic dice have to be adaptable so that they behave in the same way as the primitives and thus be able to model the behaviours of the risks.

Excel experts will point out that having a modelling tool is not essential because the functionality provided by the risk-modelling tool can be replicated entirely in standard Excel. This is because Excel has a random number generator with a uniform distribution that can be programmed to put a random integer in a designated cell with each F9 recalculation. Here is the Excel code for a one to six dice:

```
=RANDBETWEEN(1,6)
```

I can put a formula in another cell that says if the random number is 1, the adjacent cell is 1; or if it is 2 or 3, then the adjacent cell is to show 2; and if it is 4, 5 or 6 then the adjacent cell is to show 3.

```
=RANDBETWEEN(1,6)

=CHOOSE(<reference to above cell>,1,2,2,3,3,3)
```

And so with repeated F9 recalculations, the first cell will show an even distribution of ones to sixes, a uniform dice, whereas the adjacent cell will show some ones, more twos and even more threes. I have therefore transformed a flat distribution into a shaped one, a triangle, and with some further ingenuity, can make any shape needed. However, the argument against using Excel like this is that risk models can be complicated enough without adding more complexity by incorporating emulations of risk modelling algorithms based on standard Excel into the model as well.

It is much more sensible to buy a risk-modelling tool, a program that has at its heart a random number generator that can be used to emulate the behaviour of risks in ways that are easy to apply, and when studying them to see how they generated the results they did. The ease of application and assimilation provided by a risk-modelling tool is important. A great benefit of spreadsheets is to be able to replicate formulae quickly from column to column and from row to row. Switching on Excel's 'show formulae' view CTRL + ` [the grave accent key, the one usually below the Escape key] will often reveal a nicely ordered pattern of formulae laid out with guardsmen-like precision from left to right and top to bottom. If the formula had to be expanded by one more column, say if another year had to be added to a cash flow sheet, or one more row added, say for the expenditure on computer consumables, the pattern would repeat. Any break in the pattern is automatically flagged in Excel 7 with a green flash in the top-left corner of a cell, indicating a potential mistake in the orderly sequencing of the spreadsheet encoding.

This orderliness is not necessarily the case in risk models, where the formula visible in a cell is often not a reliable indicator of what is in neighbouring cells, or if a formula does repeat, the area of the sheet over which it is spread is typically small. Generally then, each patch of risk-modelling calculation will be different from those nearby, and so I tend to think of risk models as being made up of snowflakes and not the slats of a Venetian blind.

The point I wish to emphasise by the snowflakes analogy is that risk models are sufficiently complicated as they are, without incorporating the further complexities of emulations of risk-modelling algorithms using standard Excel into them as well. The dangers of complexity in spreadsheets are well known. We quickly forget how the calculations do what they do and we forget the boundaries within which their results are valid. Also, by making them more complicated we make the task of their future re-assimilation far more difficult to complete, as well as more vulnerable to misunderstandings. I have had the enervating experience of realising it would have been easier to have started again than it was to re-learn and adapt a model that had been untended and unused for two years. We can also make edits to live models, the consequences of which, in the distant unexplored corners of our Excel workbooks, we did not anticipate, and even worse, did not spot in the results. A risk-modelling tool may not keep risk modelling simple but it does keep it simpler than not using one at all.

There are several risk-modelling tools in the marketplace. My recommendation is that you buy one that is calculation-oriented, characterised by having an extensive ability to model risks in a quantitative way. These are more uncommon than those that are data-oriented. These provide a comprehensive capability to store, sort, filter, shuffle and shove out descriptions of risk information as elaborate as your imagination can conceive, provided you have entered it all in the first place and do not mind the rigidity they tend to impose even after your risk process. These tools are not rarer in the market place per se, but with the publication of the Turnbull Report on the UK Corporate Governance Code[2] in which the work by the Hampel, Cadbury and Greenbury committees on corporate governance was consolidated (and which called for directors of companies to introduce an explicit process for the management of risk), they have spread around the workplace and so are much more commonly seen than the risk-modelling tool, which usually sits on a single machine in a quiet corner. This machine can sometimes be located by the sound because whereas the data-centric ones are usually operated in silence or in neighbourly conversation, the calculation-centric ones are operated by analysts who every now and then sound like steam engines, emitting huffs, puffs, whistles and long sighs. Risk data processors always know what time it is whereas risk analysts struggle to remember because their preoccupation masks its passing. This is because though data processing is hard, tedious and diligent work, encoding and running risk models is hard, tedious and

---

2   http://www.frc.org.uk/corporate/ukcgcode.cfm, C.2 Risk Management and Internal Control. Main principle. The board is responsible for determining the nature and extent of the significant risks it is willing to take in achieving its strategic objectives. The board should maintain sound risk management and internal control systems.

diligent work that goes wrong, and so the only time a pressured analyst is often aware of is how much of it they have left.

There are many risk-modelling tools but what I need for the purposes of this book is one that provides sophisticated modelling, and which provides in such a way that it is easy to describe how it is used without reference to a user's manual or presumed attendance on a course. The type of tool I am thinking of is one that adds a set of risk modelling functions to those already provided within Excel, such that the use of them is no less familiar an operation than using the functions already provided for statistical analysis, text manipulation and so on. Tools like this are readily available and so to avoid this becoming promotional let me describe an imaginary one called RMT (Risk Modelling Tool), the use of which is typical of them all.

The best thing about RMT (I imagine) is the immediate familiarity of its look and feel. The opening screen is Excel, and so there will be no initial groan, no dent in the table from the falling forehead and no reaching for the manual as yet another user-friendly, new to you, user interface screen bursts into view. It could not be friendlier, even though in time it will become apparent that RMT has not yet really appeared on the screen. It is being kept out back in the yard while we are being welcomed into the parlour and put at our ease. Risk modelling in RMT is done in Excel and like any spreadsheet work, risk modelling makes extensive use of the formulae and functions provided by Excel. However on this occasion these are supplemented by a set of risk-modelling formulae. Thus to SUM, MIN, LEFT, LOOKUP, etc. are added risk-modelling functions such as RMT/TRIANG, RMT/MEAN, RMT/PERCENTILE, and so on.

The act of risk modelling is to choose a primitive that reflects a risk's behaviour and then to select the Excel and RMT functions that together model it appropriately. For example, say a risk model comprised a fixed number of components and a variable element: between one and six components more, on top of that. The equation for this risk would be:

Total =SUM (A1:A100) + RMT/DISCRETE ({1, 2, 3, 4, 5, 6}, {1, 1, 1, 1, 1, 1}).

The SUM is a function from standard Excel and the RMT/DISCRETE a function from RMT. Within the RMT/DISCRETE part, the {1, 1, 1, 1, 1, 1} means that each of the values 1, 2, 3, 4, 5 or 6 has an equal chance of appearing in the cell after a recalculation of the sheet, and so this part of the equation models a standard dice.

There is an important difference between the two parts. Whereas the Excel SUM function would be *completely* evaluated in a single pass through the calculation, which is usually done automatically on pressing the <RETURN> key, the RMT functions would not be.

Imagine each part of the above equation is written into its own cell. A single pass through the calculation – a single function key F9 recalculation – would be sufficient to evaluate completely.

= SUM(A1:A100)

No matter how often F9 was pressed the answer would not change.

However, a single F9 recalculation would not completely evaluate

+ RMT/DISCRETE ({1, 2, 3, 4, 5, 6}, {1, 1, 1, 1, 1, 1})

Either a 1, 2, 3, 4, 5, or 6 will show in the cell in which it has been entered. This value will probably change with every F9 (it may stay the same: a three being re-evaluated as a three, for example, on a subsequent F9 re-calculation).

Say this part of the equation had been encoded by a colleague who had then prevented the cell from displaying (in the formula bar) what they had written. We would not be able to see the RMT/DISCRETE ({1, 2, 3, 4, 5, 6}, {1, 1, 1, 1, 1, 1}), only its evaluations. We would only completely know the possible evaluations of the cell after noting the outcomes of several F9s. Even then we may be cautious in case our colleague had put a rare event into the formula, like the seven in this example which has only a one in a million chance of occurring:

+ RMT/DISCRETE ({1, 2, 3, 4, 5, 6, 7}, {1, 1, 1, 1, 1, 1, 0.000001})

We kept on pressing F9 until we felt we were confident that for all practical purposes, the number of additional components that reflected the risk on top of the fixed total of SUM(A1 : A100) was 1 to 6.

Say there were two possible groups of additional components, Type A and Type B, which needed to be added to the fixed number SUM(A1 : A100) to reflect the risk of uncertain quantity. The equation for this risk could be modelled by copying and pasting the RMT/DISCRETE part into a second cell and summing the two cells.

Risk of uncertain quantity = RMT/DISCRETE ({1, 2, 3, 4, 5, 6}, {1, 1, 1, 1, 1, 1}) + RMT/DISCRETE ({1, 2, 3, 4, 5, 6}, {1, 1, 1, 1, 1, 1})

We now have a two-dice model of the risk, and as in the simple garage project, say we need to know the minimum, maximum and most likely outturn costs of this risk. Of course we can easily see these should be 2, 12 and 7, but will the two-dice model confirm our foresight so that when we expand the modelling to a practical situation in which there could be dozens of risks, we can have confidence that when RMT calculates them for us, our experience of a simple two-dice scenario tells us they are correct.

To see if RMT confirms our foresight, we would have to throw the dice several times by pressing the F9 repeatedly, recording the sum of each recalculation. This would be laborious and un-revelatory. Thankfully, the modelling software will do the repeated calculation and calculate the statistics for us. I do not want to expand on how to operate RMT but it is easy to get RMT to do this. Typically, there will be a flurry of activity on the screen as RMT drove Excel through, say, a thousand throws of the dice, followed by what my keyboard wizard friends call a brief moment of 'wappitywop' (when windows open and close faster than the brain can follow what is happening) as Excel is sent 'out the back' and RMT takes over the screen, as often as not, for its first proper full dress appearance in the whole modelling exercise so far. Typically RMT would be displaying a histogram, a probability density function that looked like Figure 2.31.

The x-axis shows the summation found 2 to 12; and the y-axis, the number of each that occurred most often, thus 7 is the most likely, 2 is the minimum and 12 maximum. This is a trivial result but an RMT will carry out the same 'throw the dice and show the statistics' process for models that contain hundreds of dice (Figure 2.32).

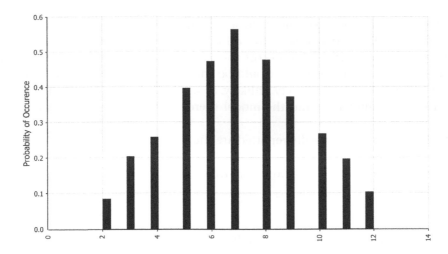

Figure 2.31 Probability density function for the garage project/two-dice model

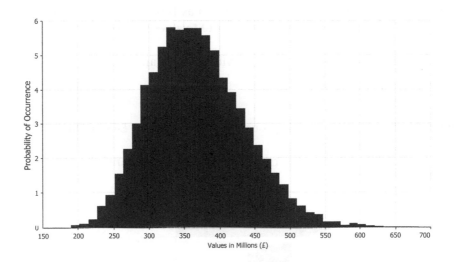

Figure 2.32 A risk exposure function for an actual project risk model containing 6000 dice and thrown 10,000 times (notice the similarity to the previous figure)

Figure 2.32 shows a similar curve to that in Figure 2.31, but in this case it is for a real project and the model that generated it contains 6000 dice. None of the dice are the uniform 1–6 type, and some of them are cross-linked so that the outcome of one is influenced in some way by the outcome of others. The RMT will still make as many throws as I choose to set, 10,000 in this case, and in time, typically two hours, automatically produce a graph like the one above that shows the minimum, maximum and most likely. Though in this case, unlike those for the two-dice example, what these may be is not obvious beforehand.

In Figure 2.32 the risk exposure is measured in £Millions and stretches from about £200M up to £600M with a peak at around £350M. In large models like this there is a convention that the exact minimum cost and the exact maximum cost are not reported. This is because the minimum cost would be the one for which every dice showed its minimum value, which is extremely unlikely, as would the situation in which every dice showed its maximum value, and the model therefore produced its maximum possible value.

In practice, the values usually reported are the ones that have

- A 95 per cent chance of not being more than – the highest.

And

- A 5 per cent chance of not being less than – the lowest.

Here is the previous figure repeated, but this time with delimiters overlaid (Figure 2.33) so that the exposure values at 5 per cent, and the 95 per cent probability, are easy to see.

If the area under curve in Figure 2.34 is taken as 100 per cent, the delimiters at 5 per cent and 95 per cent cut the curve at points at £M values where there is

- Only a 5 per cent chance the exposure will be less than £266.6M.

And

- Only a 5 per cent chance the exposure will be more than £485.61M.

By convention, this last limit is reversed to be a 95 per cent chance that exposure will not be more than £485.61M.

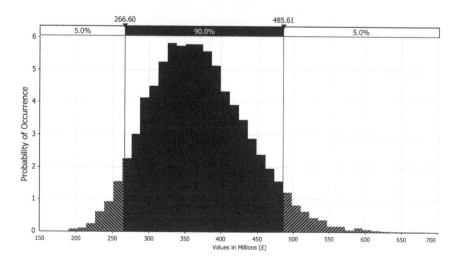

Figure 2.33 The previous figure with delimiters added at 5 per cent

These extremities, where the probabilities of the values they contain and hence the numbers shown on the faces of the dice that generated them are acceptably unlikely to occur, are shaded in the figure.

The 5 per cent and 95 per cent limits can be more easily discerned if the exposure curve is converted to cumulative form (Figure 2.34).

In cumulative form, an imaginary vertical line is swept across the exposure curve from left to right and the area to the left of it is measured. When the imaginary line is at the left end of the curve, there is no area to its left. When is at the right end, all of the area, 100 per cent, is to its left. In between, the swept area increases from left to right as the curve in Figure 46 shows.

The consequence of this is that the y-axis is transformed from being an obscure measure of absolute probability to being a linear 0–100 per cent scale. This makes reading off the 5 per cent and 95 per cent values easy to do. The drawback is that the peak value of ~£350M, and the slight skew to the right of the exposure curve in Figure 2.33, are far less easy to discern in Figure 2.34. It for this reason that most analysts prefer the probability density form of Figure 2.32. The exposure function in measured in £ and so it is a trivial matter to add the cost of the project to it to generate the overall cost function (Figure 2.35).

The figure on the left is the same exposure function from the previous figures plotted on an x-axis that has been extended to £1500M. On the right, the exposure function has been added to (in this case) the £810M project cost, plotted on the same scale. The shape of the curve does not change, it simply shifts along the x-axis. The 5 per cent and 95 per cent delimiters now refer to the total funding required for the project.

In the example I have been describing, the mid-range risk exposure of approximately £350M is 43 per cent of the estimate of £810M. The risk exposure from the risk model is therefore a critical component of the cost function and hence the project funding

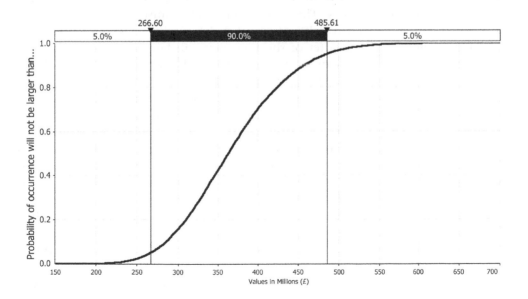

Figure 2.34 The previous figure in cumulative probability form

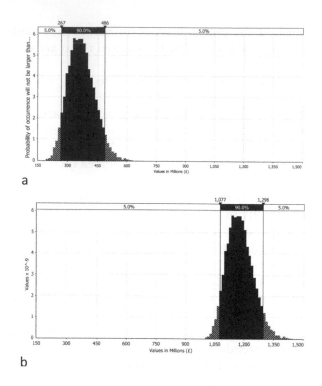

a

b

Figure 2.35 Adding the cost estimate to the exposure = the cost function

requirement. Any project given the go ahead without the risk exposure being taken into consideration when deciding how much funding it will be allocated runs a much higher chance of running out of money. The 43 per cent is not unusual. I often hear myself repeating that a well-researched risk assessment will typically justify one-third of the funding requirement, and a quarter of the time that will be needed to complete the project, though I am not advocating that projects should be given it to spend, more that project funders should make due provision in their reserves for the exposure values that emerge from the modelling. Values of the proportions emphasise the importance of the risk model in supporting an informed investment decision, as well as forewarning the analysts of the intense scrutiny to which it will be subject. I would not like to find myself subject to such examination with a na-vely modelled, under-researched and indefensible cost function, and so a question I repeatedly ask myself is: is the analysis I have done *investment grade*?, i.e. is the identification of the risk sufficiently complete and is its modelling sufficiently sophisticated so that the overall analysis is robust and reliable? I am confident that it is if the risk model has used a rich set of primitives, the risk identification exercise has found some major risks that cannot be modelled and the overall analysis has been based on extensive research, both wide and deep, into the project and its context.

Having described how a risk model is built, how it works and the important results it produces, I want to give examples of each type of primitive and set these within a demonstration risk model. The primitives are those set out on page 56 but before I do, there are two more points I must make.

1. Some of the primitives when modelled may need varying or interlinking in order to exhibit both secondary and interdependent effects. Sophisticated funders are aware of them and will scrutinise models for their presence before accepting the results as valid. A simple example: the potential of a landslip when constructing a cutting for a motorway will incur an extra cost. But it should also incur an additional delay, and when the event is deemed by the dice to have occurred in the cost side of the calculation, it should also simultaneously appear on the time side of the calculation.

2. I will be using typical quantifications in the examples because I think it makes the examples more realistic than algebra would. However, the numbers are not real, instead they are realistic. A question often arises when quantifying risks: to what degree of precision should risk quantities be modelled? My suggestion is either:

   - To a twentieth of a percentage point of the estimate, thus for a £1M project: £500, for a £1B project: £1000,000.

   Or

   - To a level at which disgrace is handed out. As a result, even on a £1B project, if some poor project manager is going to get their career blighted for being a £100,000 over budget, I would quantify to that level because that is the granularity that matters to them.

3 Risk Modelling Examples: Cost Context

The examples are a set out as a number of discrete risks together with their primitives and modelling in RMT and Excel. Together they constitute a risk model for a realistic project. Obviously I cannot set out a risk model for every type of project because no two projects are the same even when they are delivering things that are similar, and because the book would become never-ending. I have to choose as my example project a type that could be done in a context that has a wide range of parallels to other types, and whose working and technologies are hopefully familiar to everyone. I have therefore chosen a project to build a new railway.

Railway projects have the usual generic project phases:

- concept
- feasibility
- design
- construct
- commission
- operate
- maintain
- renew
- decommission and disposal

with all the finer gradations of

- outline design
- detailed design
- integrate
- test
- trial
- repair

and so on, that can be discerned within the project life cycle. Railway projects also use a wide range of engineering skills:

- civil
- electrical
- electronic
- IT and
- mechanical

and face the familiar problems of integrating them. The components of a railway asset can also be described in everyday language and so are easy to understand:

- a trackside equipment cabinet
- a signal gantry
- a twin-track tunnel

and so on.

My hope is that even for readers who do not work on railway projects the examples will be easy to understand, and with a change of terminology and technology, be useable elsewhere.

Whenever the formulae I give use large numbers, e.g. £20,000,000 I have shortened them to £20M for readability. To use what I have written in spreadsheets will require the £20M text string to be translated back into the number 20,000,000 for the formula to work.

I have taken some liberties with the exact syntax of the code in the hope that I have made it easier to read as part of the narrative flow, and I hope a not too impertinent assumption that the actual code that will make the examples work is clear.

Some of the calculations are awkward to explain in narrative form, so all of them are included in the example model spreadsheet that accompanies this book.

An Overview of the Example Model (Cost Part)

The example model, Appendix 2 An Example Risk Model, contains worked instances of the cost and time-risk primitives. Though in two parts, cost and time, this section will focus on the cost risk. It comprises approximately 90 rows of 11 columns.

In Table 3.1 the risks are described and modelled from left to right across a row from columns A to K. Columns A to D number and describe the risk, with its synopsis given in D. Column E contains an explanation of how it has been modelled, and F to H contain interim calculations; multipliers, subsidiary dice evaluations, parameters and so forth. Column I contains the current evaluation, or sample, of the risk. J and K (not shown here) contain convenient output labels for use in the post-modelling analysis described later in the book. Each risk is separated from its predecessor by a blank, blue row.

Not all of the risks occupy a single row. Many take up several rows, and some of the interim calculations are quite substantial, as shown by blocks of like-shaded cells. This is because real-world models are not Venetian blinds, by which I imply that every risk, no matter its behaviour, is modelled in the same way. This modelling is then simply clattered down a spreadsheet like a Venetian blind. I think such simplifications are the problem many clients have with risk analysis, and it's a shame that such simple, prescriptive, limited analyses generate graphs and figures every bit as elegant and beguiling as building block models do.

Table 3.1 Part of the example cost risk model

	A	B	C	D	E	F	G	H	I
1	Example risk no.	Primitive	Name	Synopsis	Modelling note	#1	#2	#3	Sample
2									
3	1	Continuous quantity	P way spread on quantity	The quantity of rail track needed is a best guess. Between −1 and +1 mile more could be required @ £1M per mile, but with reducing probability.	*Spread quantity (continuous): quantity modelled as Triangular −1.0.1*	0.09			£91,075
4									
5	2	Discrete quantity and variable price	Cabinets spread on quantity and rate	Some existing trackside equipment cabinets may have to be moved to another site. Funding for 6 has been included in the estimate, but up to 4 more may also have to be moved, depending on the final layout of the track. All numbers, from 6 to 10 10 (= 6 + 4), have an equal likelihood of happening.	*Spread quantity (discrete)*	4	4		
6				The cost of relocating a cabinet depends on the number in the subcontract, from £40k per unit down to £25k per unit.	*Spread rate (covarying)*	£25,001	£37,723		£150,893
7									

The following figure (Figure 3.1) shows a zoomed out view of a simple, Venetian blind-style model alongside one that has some complexity and bulk to it. The zoomed out view, which makes the text in the figure too small to read, is intentional and has been done to show that the model on the right on the basis of visual impact alone is instinctively closer to a state of ASANA – As Sophisticated As is Necessary to Assure – than the one on the left.

Each risk in the example model will now be extracted and described in turn.

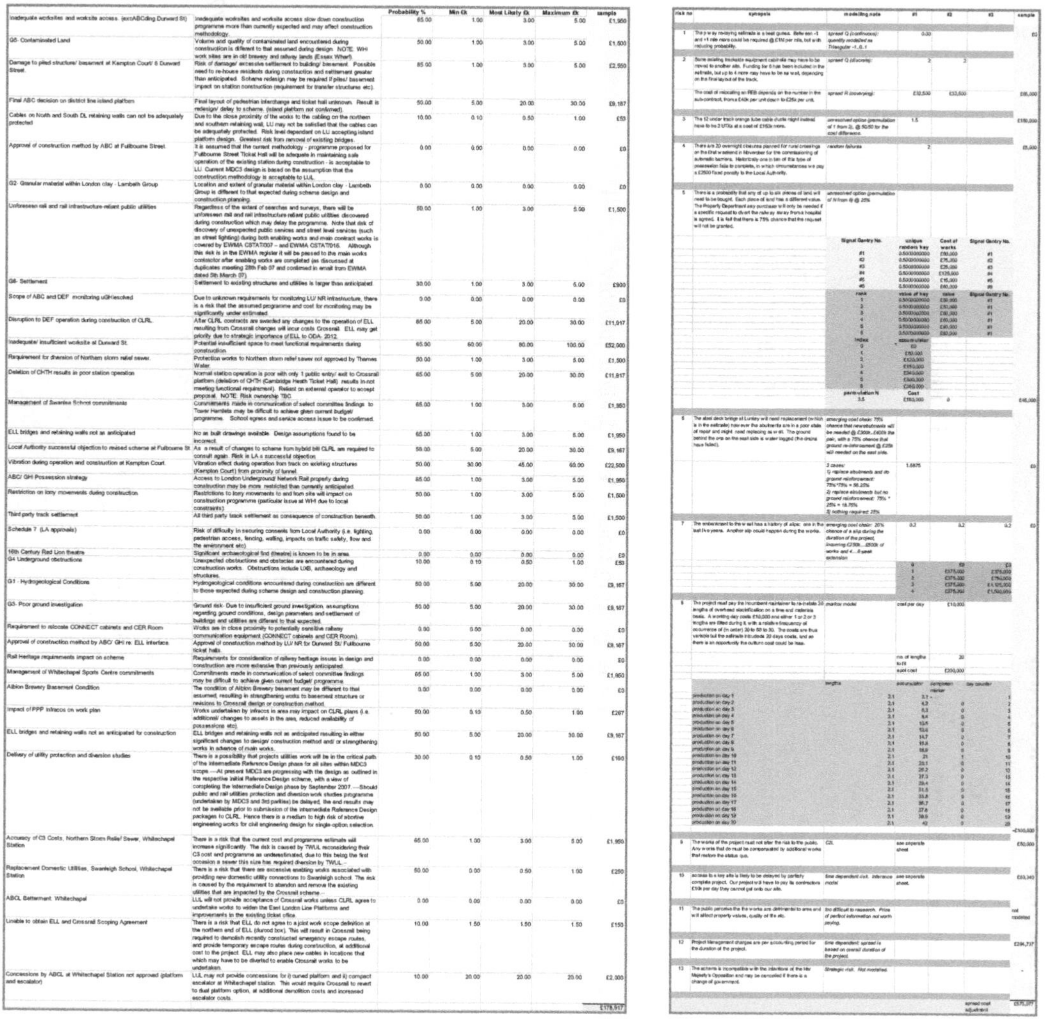

Figure 3.1   A Venetian blind model alongside a building block model

Spread on Quantity Primitive and Spread on Price Primitives

EXAMPLE RISK 1 CONTINUOUS QUANTITY

The design of the proposed new railway line is not complete, and at present the quantity of track required could be between 1 mile less and 1 mile more than the quantity used in the estimate.

• See cells A3: I3 in the cost risk modelling sheet of the example model (Table 3.2).

Table 3.2   Example of spread on quantity – continuous

Row/ column	D	E	F	G	H	I
	Synopsis	Modelling note	#1	#2	#3	Sample
3	The quantity of railway track is a best guess. Between –1 and +1 mile more could be required @ £1M per mile, but with reducing probability.	Spread Q (continuous): quantity modelled as triangular –1.0.1	= RMT/ Triang (–1,0,1)			= D3 * 1000000

This risk has arisen because both the route of the railway line and the layout in terms of the number of sidings and loops has not yet been finalised (Figure 3.2). It is a risk because of a future situation, which may cause an overspend. Equally though it is an opportunity for an underspend, but as discussed, I will talk about both using the term 'risk'.

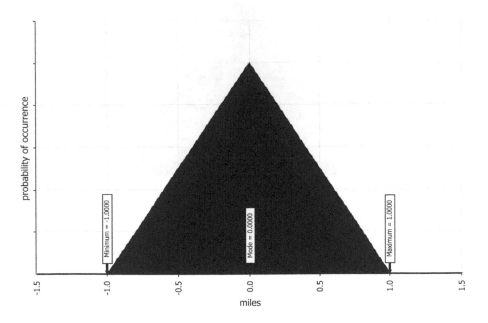

Figure 3.2   The risk on the quantity of railway track needed

The risk may match similar risks elsewhere; cabling in the IT sector, pipe work in the petrochemical sector, ducting in the building sector and so on.

Suppose the analyst, in discussion with the designers, finds that the probability of occurrence of the value of the quantity is increasingly unlikely the more it is, so that Figure 3.3, represents the behaviour of this risk. Further from discussion with the estimators, the analyst finds that track costs a £1M per mile. The risk exposure is thus (Figure 3.3).

The formula for the risk is:

= RMT/Triang (–1,0,+1) * £1M

- see cells F3 and I3

where the RMT/Triang function from RMT is used to generate a random number anywhere between –1 and +1, with the value most frequently generated being 0, and values out to either + or –1 being increasingly rarer.

If there was no strong view about the existence of a declining probability so that all values between –1 mile and +1 mile may equally occur then the analyst may decide that

= RMT/Uniform (–1,+1) * £1M

would be a better model (Figure 3.4).

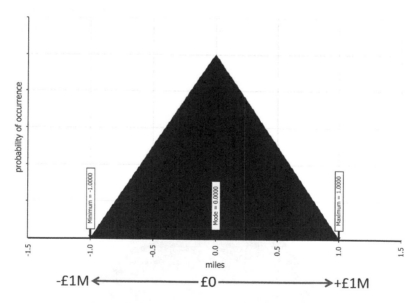

Figure 3.3   The risk exposure on the quantity of railway track

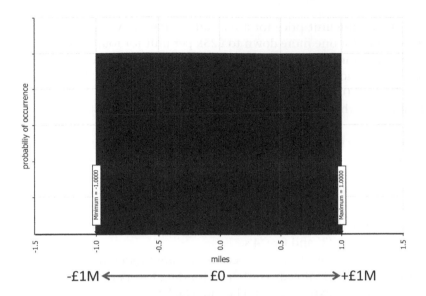

Figure 3.4   The risk exposure on the quantity of railway track if all possible values are equally likely, i.e. there is no opinion about which values are more likely than others

EXAMPLE RISK 2: DISCRETE QUANTITY AND VARIABLE PRICE

Some existing trackside equipment cabinets may have to be moved to another site. Funding to move six has been included in the estimate, but up to four more may also have to be moved, depending on the final layout of the track. All numbers, from 6 to 10 (= 6 + 4), have an equal likelihood of happening.

• See cells A5 : I5 in the cost risk modelling sheet of the example model (Table 3.3).

Table 3.3   Example of spread on quantity – discrete

Row/ column	D	E	F	G	H	I
	Synopsis	Modelling note	#1	#2	#3	Sample
5	Some existing trackside equipment cabinets may have to be moved to another site. Funding for 6 has been included in the estimate, but up to 4 more may also have to be moved, depending on the final layout of the track. All numbers, from 6 to 10 10 (= 6 + 4), have an equal likelihood of happening.	Spread Q (discrete):	= RMT/Discrete ({0,1,2,3,4}, {1,1,1,1,1})			

The estimators say the unit price for a relocation changes with the number instructed, from £40k per unit for one more down to £25k per unit for four more.

This too is an example of a spread on QUANTITIES Primitive but with an important difference – there is no such thing as 1.2 cabinets or 3.5 cabinets. There has to be a whole number, in this case 0, 1, 2, 3 or 4 (Figure 3.5).

The formula for the quantity part of the risk is:

= RMT/Discrete ({0,1,2,3,4}, {1,1,1,1,1})

See cell F5, where the RMT discrete function is used to model the extra number of cabinets. The {1,1,1,1,1} is RMT code that sets the relative probabilities of the {0,1,2,3,4}, which because they are all '1' are therefore equal. They could be set differently, for example {1,1,1,1,5} would say that on average, for every 9 throws of the dice there would be one '0', one '1', one '2', one '3' and five '4's.

Note the possibility of '0'. This is because we are modelling the exposure over and above the estimate, which already contained funding for six cabinets.

The cost part of the exposure could be modelled as:

= RMT/Uniform (£25k,£40k)

Giving an overall exposure of:

= RMT/Discrete ({0,1,2,3,4}, {1,1,1,1,1}) * RMT/Uniform (£25k,£40k)

But this would be wrong because the unit price depends on the number of extra cabinets that need to be moved, and if the number of cabinets goes up from one throw of the dice to the next (for example, throw 1 = 2, and then throw 2 = 4) then the unit price must go down. Alternatively, if the number goes down then the price must go up. A scenario in which the number and the price both went up or both went down would be wrong,

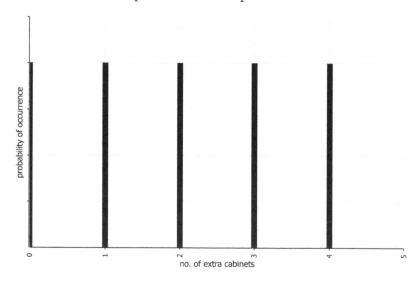

Figure 3.5   Spread on the number of cabinets

and that is what may well happen with the exposure formula (above) when the dice are thrown.

The solution is to introduce covariance. This takes the form of additional coding that ensures the RMT formulas depend on each other in a specific way. In this is case, if one goes up the other must come down.

Here is the actual modelling:

Number = RMT/IndependentVariable ('aaa') + RMT/Discrete ({0,1,2,3,4}, {1,1,1,1,1})

Price = RMT/DependentVariable ('aaa',–1) + RMT/Uniform (£25k,£40k)

See cells A5 : E6 in the cost risk modelling sheet of the example model (Table 3.4).

The number in cell G5, is appended with an identifying label 'aaa' and marked by the coding 'indepc' to be an independent variable. Its value can move up or down from dice throw to dice throw.

The price however may not. It has to respond to movements in 'aaa', the number, in the manner set by the value –1 because it has been encoded as a dependent variable, 'depc'. The –1 says it must decrease when the number increases and vice versa.

The formula for the risk is thus the product of the two formulae:

= RMT/Discrete ({0,1,2,3,4}, {1,1,1,1,1}, RMT/IndependentVariable ('aaa')) * RMT/ Uniform (25000,40000, RMT/DependentVariable ('aaa',–1))

The parameter –1 is the covariance coefficient. If it were made +1, the number and the price would move in the same direction, increase or decrease, on the same throw; –1 forces them to move in opposite directions; when the independent variable goes up, the

Table 3.4   Example of covarying spreads on quantity and rate

Row/ column	D	E	F	G	H	I
	Synopsis	Modelling note	#1	#2	#3	Sample
5	Some existing trackside equipment cabinets may have to be moved to another site. Funding for 6 has been included in the estimate, but up to 4 more may also have to be moved, depending on the final layout of the track.	Spread Q (discrete):	= RMT/Discrete ({0,1,2,3,4}, {1,1,1,1,1})	= RMT/Discrete ({0,1,2,3,4}, {1,1,1,1,1}, RMT/ Independentvariable ('aaa'))		
6	The cost of relocating an REB depends on the number in the subcontract, from £40k per unit down to £25k per unit.	Spread R (covarying):	= RMT/ Uniform (25000,40000)	= RMT/Uniform (25000,40000, RMT/ Dependentvariable ('aaa',–1))		= E5*E6

dependent one goes down, and vice versa. Other values between −1 and +1 are allowed for cases where the covariance is not absolutely together or absolutely not together, but instead reflective the tendency to behave in the desired direction.

Digression: So Far ...

So far I have modelled two risks, the first an uncomplicated use of the spread on quantity primitive, and the second a combination of the spread on quantity primitive in a different form (or 'colour' as I termed it earlier in the book) and a spread on prices that was made necessarily more sophisticated by adapting it to covary with quantity.

Even with these two elementary risks, elementary in the sense that they are easy to identify and to understand, a degree of necessary sophistication has emerged in their modelling without which their validity would be questionable.

If coding for both risks were entered into a RMT/Excel spreadsheet we would then have a simple risk model that we could instruct RMT to run over 5000 (say) iterations. On each iteration RMT will choose a new random value for every RMT function in the model, according to the range specified in the parameters, ensuring that over the 5000 iterations values appear in accordance with the specific or implicit probability of their modelling.

If we were to sum the outputs of the two risks, and after the 5000 iterations plot a distribution of the values of the sum, this would be the exposure curve. Already the picture that may have been expected is emerging. Note how the opportunity for there to be one mile less of railway track shows the exposure can be negative, which is to say there is an opportunity here for cost saving. If we were to add the estimate into this curve we would get the cost function (Figure 3.6).

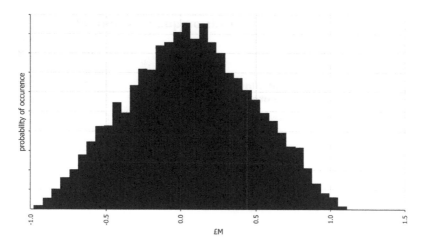

Figure 3.6   The exposure curve – two risks so far

EXAMPLE RISK 3: VARIABLE QUANTITY BASED ON PRODUCTIVITY

The project must pay the incumbent maintainer to reinstate 20 lengths of overhead electrification on a time and materials basis. A working day costs £10,000 and 1, 2 or 3 lengths are fitted during it, with a relative frequency of occurrence of (in order) 20 to 50 to 30. The outturn costs are thus variable, but the estimate includes 20 days costs so there is opportunity possibility the outturn cost could be less.

- See cells A8 : I33 in the cost risk modelling sheet of the example model (Table 3.5).

This example contains a substantial internal calculation table in cells D58 : H78 in which the production time for each of twenty days: one or two or three units is determined by the formula:

= RMT/Discrete ({1, 2, 3}, {20, 50, 30})

The risk is actually an opportunity because the estimate, by containing funds for 20 days work, implicitly assumes only the lowest productivity will be achieved, whereas the historical record shows this could be twice or even three times as much.

A day by day total is accumulated in column E, and a logical test is made in column F – how many days it took to complete the total of 20 units. This is marked by a 1 in column G. A lookup function in cell H79 finds the number of days taken and computes the difference between the estimate cost and the modelled cost.

A similar example appears in the time-risk model, in which day by day improvements in productivity rates are used to predict the amount of time needed to complete task. If that time were less than estimated, and the production team were on time-hire then a link could be made back to this model to predict further cost savings (see Example Risk 27).

Unresolved Option Primitive

EXAMPLE RISK 4: ONE OF TWO OPTIONS

The designers have not yet decided how to get a set of cables from one side of the tracks to the other. The estimators have assumed plastic tube ducts will be used but the designers may opt for a more expensive concrete conduit at a cost of £150k more.

- See cells A35 : I35 in the cost risk modelling sheet of the example model (Table 3.6).

Here the analyst has discovered that the designers have more than one solution to an engineering problem in mind. It could easily be a procurement team that has yet to decide between two quotations for the supply of parts, or a construction team that have not decided whether to build from the middle outwards or the outside inwards.

Table 3.5 Example of variable production using historical rates

Row/column	D	E	F	G	H	I
8	**Synopsis**	**Modelling note**	Cost per day	£10,000		
	The project must pay the incumbent maintainer to reinstate 20 lengths of overhead electrification on a time and materials basis. A working day costs £10,000 and either 1, 2 or 3 lengths are fitted during it, with a relative frequency of occurrence of (in order) 20 to 50 to 30. The outturn costs are thus variable but the estimate includes 20 days' costs, so there is a possibility the outturn cost could be less.					
9			No. of lengths to fit	20		
10			spot cost	= F55*F54		
11						
12		Lengths	Accumulator	Completion marker	Day counter	
13	Production on day 1	= RMT/discrete ({1,2,3}, {20,50,30})	= D59	–	1	
14	Production on day 2	= RMT/Discrete ({1,2,3}, {20,50,30})	= E59+D60	= if (AND(E59<$F$55, E60> = $F$55),1,0)	2	
15	Production on day 3	= RMT/Discrete ({1,2,3}, {20,50,30})	= E60+D61	= if (AND(E60<$F$55, E61> = $F$55),1,0)	3	
16	Production on day 4	= RMT/Discrete ({1,2,3}, {20,50,30})	= E61+D62	= if (AND(E61<$F$55, E62> = $F$55),1,0)	4	
17	Production on day 5	= RMT/Discrete ({1,2,3}, {20,50,30})	= E62+D63	= if (AND(E62<$F$55, E63> = $F$55),1,0)	5	
18	Production on day 6	= RMT/Discrete ({1,2,3}, {20,50,30})	= E63+D64	= if (AND(E63<$F$55, E64> = $F$55),1,0)	6	
19	Production on day 7	= RMT/Discrete ({1,2,3}, {20,50,30})	= E64+D65	= if (AND(E64<$F$55, E65> = $F$55),1,0)	7	

Table 3.5 *Concluded*

Row/column	D	E	F	G	H	I
	Synopsis	Modelling note				
20	Production on day 8	= RMT/Discrete ({1,2,3}, {20,50,30})	= E65+D66	= if (AND(E65<$F$55, E66> = $F$55),1,0)	8	
21	Production on day 9	= RMT/Discrete ({1,2,3}, {20,50,30})	= E66+D67	= if (AND(E66<$F$55, E67> = $F$55),1,0)	9	
22	Production on day 10	= RMT/Discrete ({1,2,3}, {20,50,30})	= E67+D68	= if (AND(E67<$F$55, E68> = $F$55),1,0)	10	
23	Production on day 11	= RMT/Discrete ({1,2,3}, {20,50,30})	= E68+D69	= if (AND(E68<$F$55, E69> = $F$55),1,0)	11	
24	Production on day 12	= RMT/Discrete ({1,2,3}, {20,50,30})	= E69+D70	= if (AND(E69<$F$55, E70> = $F$55),1,0)	12	
25	Production on day 13	= RMT/Discrete ({1,2,3}, {20,50,30})	= E70+D71	= if (AND(E70<$F$55, E71> = $F$55),1,0)	13	
26	Production on day 14	= RMT/Discrete ({1,2,3}, {20,50,30})	= E71+D72	= if (AND(E71<$F$55, E72> = $F$55),1,0)	14	
27	Production on day 15	= RMT/Discrete ({1,2,3}, {20,50,30})	= E72+D73	= if (AND(E72<$F$55, E73> = $F$55),1,0)	15	
28	Production on day 16	= RMT/Discrete ({1,2,3}, {20,50,30})	= E73+D74	= if (AND(E73<$F$55, E74> = $F$55),1,0)	16	
29	Production on day 17	= RMT/Discrete ({1,2,3}, {20,50,30})	= E74+D75	= if (AND(E74<$F$55, E75> = $F$55),1,0)	17	
30	Production on day 18	= RMT/Discrete ({1,2,3}, {20,50,30})	= E75+D76	= if (AND(E75<$F$55, E76> = $F$55),1,0)	18	
31	Production on day 19	= RMT/Discrete ({1,2,3}, {20,50,30})	= E76+D77	= if (AND(E76<$F$55, E77> = $F$55),1,0)	19	
32	Production on day 20	= RMT/Discrete ({1,2,3}, {20,50,30})	= E77+D78	= if (AND(E77<$F$55, E78> = $F$55),1,0)	20	
33						= (VLOOKUP (1,F60:G78,2,FALSE) * F54) − F56

Table 3.6   Example of unresolved option – 1 from 2

Row/column	D	E	F	G	H	I
	Synopsis	Modelling note	#1	#2	#3	Sample
35	The 12 under-track plastic tube cable ducts might instead have to be 2 under-track crossings at a cost of £150k more.	Unresolved option (permutation of 1 from 2), @ 50/50 for the cost difference.	= RMT/ Discrete ({1,0}, {50,50})			= D8 * 150000

The formula for the risk is

= RMT/Discrete ({1,0}, {1,1}) * £150k

• See cells F35 and I35.

Here the RMT/Discrete function used previously to model 1 to 6 and 1 to 5 is used to model 0 or 1 with an equal probability of it being either, i.e. a coin. A 0 will nullify the £150k, while a 1 will add £150k to the total of the other risks in the model.

This is clearly a trivial example, but it can be expanded to include two or more options by adding some Excel code

= RMT/Discrete ({1,2,3}, {1,1,1})

= choose (<reference to the above>, £100k, £150k, £200k)

However, the unresolved option can often be much more complicated, as in the following example.

EXAMPLE RISK 5: N FROM M OPTIONS

There is a probability that any of up to six pieces of land will need to be bought. Each piece of land has a different value. The Property Department say purchase will only be needed if a specific request to divert the railway away from a hospital is agreed. It is felt that there is 75 per cent chance that the request will not be granted.

• See cells A37 : I61 in the cost risk modelling sheet of the example model (Table 3.7).

Suppose the analyst, when researching this risk, realises that if only one piece of land needs to be bought, it could be any one of the six. And if two, then any two from the six. Only if six are bought can there be any certainty about which prices to include in the model.

This is quite a common risk-modelling scenario in the concept and feasibility stages of a project, when different subsets of all the things that could be done or bought for the project are thought sufficient; or the corollary: of all the things that are thought necessary to do or to buy, some subsets may be found not to be needed.

Table 3.7 Example of unresolved option – 1 from N

Row/column	D Synopsis	C Modelling note	D #1	E #2	F #3	G Sample
37	There is a probability that any of up to six pieces of land will need to be bought. Each piece of land has a different value. The Property Department say purchase will only be needed if a specific request to divert the railway away from a site of scientific interest is agreed. It is felt that there is 75% chance that the request will not be the case.	Unresolved option (permutation of N from 6) @ 25%				
38			Unique random key	Cost of land parcel	Signal gantry no.	
39			#1 = ROUND (RMT/Uniform (0,1),10)	= RMT/Uniform (20000,100000)	#1	
40			#2 = ROUND (RMT/Uniform (0,1),10)	= RMT/Uniform (50000,100000)	#2	
41			#3 = ROUND (RMT/Uniform (0,1),10)	= RMT/Uniform (20000,30000)	#3	
42			#4 = ROUND (RMT/Uniform (0,1),10)	= RMT/Uniform (100000,150000)	#4	
43			#5 = ROUND (RMT/Uniform (0,1),10)	= RMT/Uniform (10000,20000)	#5	
44			#6 = ROUND (RMT/Uniform (0,1),10)	= RMT/Uniform (20000,100000)	#6	
45		Rank	Value of key	Value	Signal gantry no.	
46		1	= LARGE ($D$4:$D$9,C11)	= VLOOKUP (D11,$D$4:$E$9,2,FALSE)	= VLOOKUP (D11,$D$4:$F$9,3,FALSE)	
47		2	= LARGE ($D$4:$D$9,C12)	= VLOOKUP (D12,$D$4:$E$9,2,FALSE)	= VLOOKUP (D12,$D$4:$F$9,3,FALSE)	

Table 3.7   Concluded

Row/column	Synopsis	Modelling note	#1	#2	#3	Sample
48		3	= LARGE ($D$4:$D$9,C13)	= VLOOKUP (D13,$D$4:$E$9,2,FALSE)	= VLOOKUP (D13,$D$4:$F$9,3,FALSE)	
49		4	= LARGE ($D$4:$D$9,C14)	= VLOOKUP (D14,$D$4:$E$9,2,FALSE)	= VLOOKUP(D14,$D$4:$F$9,3,FALSE)	
50		5	= LARGE ($D$4:$D$9,C15)	= VLOOKUP(D15,$D$4:$E$9,2,FALSE)	= VLOOKUP(D15,$D$4:$F$9,3,FALSE)	
51		6	= LARGE ($D$4:$D$9,C16)	= VLOOKUP (D16,$D$4:$E$9,2,FALSE)	= VLOOKUP (D16,$D$4:$F$9,3,FALSE)	
52		Index	Accumulator			
53		0	0			
54		1	= SUM ($E$11:E11)			
55		2	= SUM ($E$11:E12)			
56		3	= SUM ($E$11:E13)			
57		4	= SUM ($E$11:E14)			
58		5	= SUM ($E$11:E15)			
59		6	= SUM ($E$11:E16)			
60		Permutation N	Cost			
61		= RMT/Discrete ({1,2,3,4,5,6}, {1,1,1,1,1,1})	= VLOOKUP (C26,C18:D24,2)	= RMT/Discrete ({1,0}, {25,75})		= D26 * E26

The formula for this risk is complicated, but I want to include it because as well as computing the exposure, it also shows clearly that risk calculations are not always a few cells – single row construction, but can be big.

Each of the six pieces of land is judged by the property department (presumably) to have a minimum to maximum cost. In a second column, a random number between 1 and 100,000,000. I will refer to this as the key. These are set out in what I will refer to as the first table.

No. 1 = RMT/Uniform (£20k, £100k), = RMT/Uniform (1, 100000000)

No. 2 = RMT/Uniform (£50k, £100k), = RMT/Uniform (1, 100000000)

No. 3 = RMT/Uniform (£20k, £30k), = RMT/Uniform (1, 100000000)

No. 4 = RMT/Uniform (£100k, £150k), = RMT/Uniform (1, 100000000)

No. 5 = RMT/Uniform (£10k, £20k), = RMT/Uniform (1, 100000000)

No. 6 = RMT/Uniform (£20k, £100k), = RMT/Uniform (1, 100000000)

When executed in the model, the cells in the first column will flash up possible prices for the pieces within the ranges set, and in the second a random key.

I assume that the range of values available to each random key, 1 to 100,000,000, is sufficiently large for no two keys to have the same value on a single iteration of the model, aka a single throw of the dice.

A second table in the spreadsheet uses Excel's LARGE function to find the largest key in the second column of the first table, and using Excel's LOOKUP functions, populates itself with the identifying number of the piece of land and its random £price.

The LARGE function can be used to repeat the search and copy algorithm for the second-largest random key and so on until the sixth.

Since the keys all change with each iteration of the model, the search and copy algorithm is also repeated. The effect is to shuffle a deck of six cards, onto each of which has been written the identifying number of a piece of land and its current random but within range £price.

In a six-row column alongside the second table is kept an accumulated total algorithm, the top cell of which contains the value of the piece of land The second cell shows the total value of the top two pieces of land, and the third cell the top three and so on.

With each iteration of the model, the piece of land that is at the top, in second place, in third place and so on is determined by their random keys. This order changes every iteration.

A RMT/Discrete function determines how many pieces of land may need to be bought. This too changes from iteration to iteration:

= RMT/Discrete ({1,2,3,4,5,6}, {1,1,1,1,1,1})

This number is used to index the accumulated total of the land prices. If the index is, say, three then the total price for three pieces of land is the output, but which three they are is randomly set by the six-card shuffle routine. If on the next iteration the index was three again, then more than likely it would be a different three pieces of land at the top of the shuffle. Of course the Index could be 1, 2, 3, 4, 5 or 6 on that iteration.

Whichever accumulated £price the index points to is the initial output from the model. The final output is determined by the 75 per cent probability that this property purchase will not proceed.

= RMT/Discrete ({1,0}, {25,75}) * <accumulated price>

This is a difficult calculation to follow in narrative form. Along with all the others, it is included in the example model spreadsheet that accompanies this book.

Emerging Cost Chain Primitive

EXAMPLE RISK 6: ONE-TIME COST

The steel deck bridge at Lumley will need replacing (which is in the estimate), however the abutments are in a poor state of repair and may need replacing as well. The ground behind the one on the east side is waterlogged (the drains have failed).

- See cells A63 : I64 in the cost risk modelling sheet of the example model (Table 3.8).

Table 3.8   Example of emerging cost chain

Row/ column	D	E	F	G	H	I
	Synopsis	Modelling note	#1	#2	#3	Sample
63	The steel deck bridge at Lumley will need replacement (the cost of which is in the estimate) however the abutments are in a poor state of repair and might need replacement also. The ground behind the one on the east side is waterlogged because the drains have failed and may need reinforcement.	Emerging cost chain: 75% chance that new abutments will be needed @ £300–£400k the pair, with a 75% chance that ground reinforcement @ £25k will needed on the east side.				
64		3 cases: 1) replace abutments and do ground reinforcement: 75%*75% = 56.25% 2) replace abutments but no ground reinforcement: 75% * 25% = 18.75% 3) nothing required: 25%	= RMT/Discrete ({1,2,3}, {56.25,18.75,25})			= IF (D3 = 1, RMT/Uniform (300000,400000,+ 25000, IF(D3 = 2, RMT/Uniform (300000,400000),0))

This is the same risk as described in the overview of the risk primitives (Figure 3.7).

There are three possible ways in which the work may unfold. There might be others of course but let me assume the analyst's researchers concluded that these are the rational ones (Figure 3.8).

There is a possibility that the works will go as planned. The costs for this are in the estimate. However, there is a possibility that new abutments will be needed, and if they are then there is a chance the drainage repairs will be needed as well.

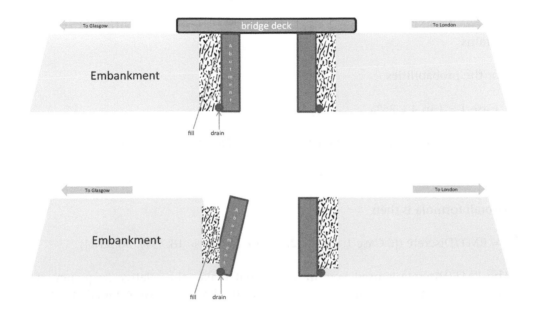

Figure 3.7   A sketch of the bridge works and the unwanted event

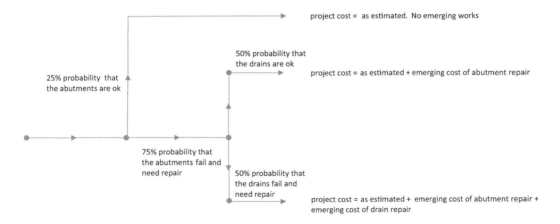

Figure 3.8   The emerging cost chain

If the probability of the works proceeding as planned is 1 in 4, 25 per cent, and the probability of the drains needing repair is 3 in 4, 75 per cent;[1] and the cost of the abutments is between £300k and £400k and the drains, £25k, the formulae for the cost outcomes of the risk are

£ Case 1 = £0 – no additional work needs to be done.

£ Case 2 = RMT/Uniform (£300k, £400k) – repair the abutments

£ Case 3 = RMT/Uniform (£300k, £400k) + £25k – repair the abutments and fix the drains

And for the probabilities

Case 1 = 1 in 4 = 25%

Case 2 = 3 in 4, 75%, and then 1 in 4, 25% = 75% * 25% = 18.75%

Case 3 = 3 in 4, 75%, and then 3 in 4, 75% = 75% * 75% = 56.25%

The overall formula is then

= RMT/Discrete ({£ Case 1, £ Case 2, £ Case 3}, {25%, 18.75%, 56.25%})

The RMT/Discrete is used here in a way that makes the relative frequency of the outcomes of the model match those expected of the risk, as if it were a loaded dice.

EXAMPLE RISK 7 MULTIPLE COST WITH TIME CROSS-LINK

The embankment to the west has a history of slips with one every five years on average. The works are planned to take a year and so there is a 20 per cent chance of a slip during the duration of the project. This would incur £250–500k of emerging works and take something like 4–6 weeks to complete.

- See cells A66 : I75 in the cost risk modelling sheet of the example model (Table 3.9).

The cost elements of this risk are *not* modelled by

= RMT/Discrete ({1,0}, {20%,80%} * RMT/Uniform (£250k, £500k),

- See cell F67 above.

---

1   The reader might question if these probabilities can be reliably known. The honest answer is yes, within reason. Most works that may be affected by the unwanted outcomes of risks will have been done before and many of the works will have been carried out by people who have been around for years. Provided the probabilities are kept broad, say in bounded approximate quartiles: never/1–25%/26–50%/51–75%/76–99%/always, then experienced people will have little difficulty in choosing and agreeing within which band the risk should sit. It would then be reasonable for the analyst to choose a mid-band probability for the modelling, or if reaching for ASANA, to model the probability as a range viz: X% = RMT/Uniform (51%, 75%), then Event occurring = RMT/Binomial (1, X%).

Table 3.9 Example of emerging cost chain with cross link to time model

Row/ column	D Synopsis	E Modelling note	F #1	G #2	H #3	I Sample
66	The embankment to the west has a history of slips: one in the last five years. Another slip may happen during the works.	Emerging cost chain: 20% chance of a slip during the duration of the project, incurring £250–£500k of works and 4–6 week extension				
67		Landslip occurs?	= RMT/Discrete ({1,0}, {20,80})			
68		No. of landslips	= RMT/Poisson (0.2)			
69		Constraint on no. of landslips	= IF(E2>4,4,E2)			
70						
71		No landslip	0	0	0	
72		First landslip	1	= RMT/Uniform (250000,500000)	= SUM($E$3:E4)	
73		Second landslip	2	= RMT/Uniform (250000,500000)	= SUM($E$3:E5)	
74		Third landslip	3	= RMT/Uniform (250000,500000)	= SUM($E$3:E6)	
75		Fourth landslip	4	= RMT/Uniform (250000,500000)	= SUM($E$3:E7)	= VLOOKUP (F69,F71:H75,3,false)

Setting the parameters of the RMT/Discrete function to a 20 per cent chance that something will happen and an 80 per cent chance that it will not, gives a 1 in 5 chance of an event that will cost something between £250k and £500k, because it contains two mistakes.

First – the average incidence of landslips is one in five years, which does not mean there is one landslip every five years. There could be several, followed by a period of many years in which none happened, but so that overall the average is one in five years. Therefore we have to allow for more than one slip to occur within the year the project is on site, because it might be a bad year.

The correct formula is

= RMT/Poisson (20%)

• See cell F68 above.

This will generate from iteration to iteration the integer number of landslips that may occur in the year in question. This will mainly be zero, occasionally one, every now and again two, rarely three and so on (Figure 3.9).

RMT/Poisson models the possibility of a bad year, whereas with the incorrect RMT/Discrete formula, only one or none would ever be reflected in the risk exposure.

Secondly – though the costs in this case would be correctly modelled by the formula

= RMT/Uniform (£250k, £500k)

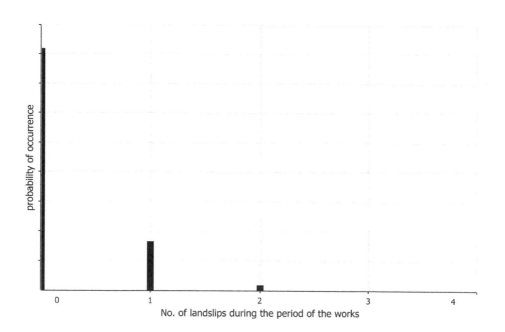

Figure 3.9   Number. of landslips during the one-year project, generated by a Poisson (20%) function

.computing the product of this and the RMT/Poisson as the risk exposure would be incorrect. In the case of more than a single landslip occurring (2, 3…), they would all have the same cost, and if we were to propagate the error into the time part of the model, take the same number of weeks to remedy.

Therefore, as so often in risk modelling, the correct code is bulky and populated within interim calculations. First, the number of slips needs to be constrained as follows

No. of land slips = RMT/Poisson (20%)

If number of landslips >4 then force number of landslips = 4

- See cell F69.

Limiting the number of landslips to four is a pragmatic step taken to prevent the model having to be expanded needlessly to accommodate increasingly rare situations in which the Poisson function generates five or more landslips. Five or more landslips would be an exceptionally rare event for a RMT/Poisson at 20 per cent, about 1 or 2 instances in 100,000 occasions, i.e. if the project were to be done 100,000 times, on one or two occasions, five landslips would occur. There is little point in expanding the model to include computations that will for all intents and purposes not be used.

Next, the consequences for each slip in terms of cost and time need to be evaluated independently.

Cost of landslip 1 = risk uniform (£250k, £500)

Time to remedy landslip 1 = RMT/Uniform (4,6) weeks

Cost of landslip 2 = risk uniform (£250k, £500)

Time to remedy landslip 2 = RMT/Uniform (4,6) weeks

Total cost so far, 2 landslips = accumulated costs of the above two

Total time so far, 2 landslips = accumulated weeks of the above two.

- See cells G71 to H75 for the costs.

The evaluations of the times are in the time risk model which is described in the next section. Note the possibility of zero landslips happening: cells D42 to F42 … and so on, out to landslip five, accumulating the cost and time for each instance.

The value generated by the RMT/Poisson (after constraining to four in this example) is then used to read off the cost required for that number of landslips using a standard Excel lookup function (see I75).

The cross-link to the time risk model will be covered in the next section.

Random Failures Primitive

EXAMPLE RISK 8: RANDOM FAILURE

There are 20 (combined) overnight road and rail closures planned for rural road/railway level crossings on the first weekend in November for the commissioning of automatic barriers. Historically, one in ten occurrences of this type of weekend work fails to be completed on time, either because of problems emerging on site or because of the late arrival of key staff. If a crossing is not reopened on time then a fixed penalty of £2500 is usually paid to the local authority.

- See cells A77 : I77 in the cost risk modelling sheet of the example model (Table 3.10).

A similar situation to the Poisson solution of Example 6 pertains here as well. That one in ten works fail to complete is an average that does not imply that here. With 20 works to do, two will fail and so the exposure is a certain 2 * £2500.

Consider a coin. Just because one in two is the average a head appears does not mean that throwing five heads in succession is impossible, just that it is rare. It is the same here; that ten instead of two sites fail to complete is not impossible but simply rare. On the other hand, four sites failing may well be quite a common outcome, as may none. We therefore need a function that usually generates the number of failures expected, but occasionally generates more or fewer such that the overall average remains the number expected. Further, numbers of failures that are increasingly remote from the expected value are increasingly unlikely. The function needed is the binomial, and the formula for the number of sites failing to finish is = RMT/binomial (20, 10%) * £2500

Table 3.10   Example of random failures

Row/ column	D	E	F	G	H	I
	Synopsis	Modelling note	#1	#2	#3	Sample
77	20 overnight closures are planned for rural crossings for the commissioning of automatic barriers. Historically one in ten of this type of possession fails to complete, in which circumstances a £2500 fixed penalty has to be paid to the local authority.	Random failures	= Riskbinomial (20,10%)			= D2 * 2500

The Riskbinomial function will generate an integer number (Figure 3.10)
This primitive is another variation of the spread on quantity primitive.

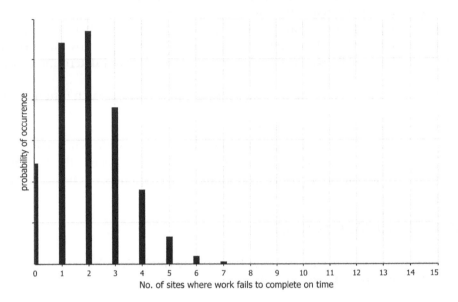

Figure 3.10 Number of sites failing, generated by binomial (20, 10%)

Not Modelled Primitive

EXAMPLE RISK: 9 PRICE OF PERFECT INFORMATION

The public perceive the works to be detrimental to the area, and that it will affect property values, increase traffic congestion and reduce trade in the town centre etc.

- See cells A79 : I79 in the cost risk modelling sheet of the example model (Table 3.11).

Table 3.11   Example of the price of perfect information

Row/ column	D	E	F	G	H	I
	Synopsis	Modelling note	#1	#2	#3	Sample
79	The public perceive the works are detrimental to area and will affect property values, quality of life etc.	Too difficult to research. Price of perfect information not worth paying.				Not modelled

All of the discrete exposures implicit in this risk are quantifiable, but only through what would probably be an extensive programme of research within the community. The programme may even heighten expectation of windfall profits to be made, thus altering the quantifications of the risk. The resulting model would probably be a large one to encode and its input data difficult to keep credible, particularly when other economic influences that may affect the same measures of well-being (job losses) also vary at the same time. The results would therefore be questionable and determined as not of investment quality.

One way to improve them might be to commission further research into normalising out the other economic influences. While this may well be a good thing to do, I think it is a line of inquiry more for academe to pursue and not analysts developing a business case. I would frankly suggest that the price of having perfect information is not worth paying. It may be in terms of the cost of its research (often it is not), but generally the timescales needed are more than the project can accommodate. This risk needs to be declared as not modelled.

EXAMPLE RISK 10: STRATEGIC RISK

The scheme is incompatible with the intentions of the Her Majesty's Opposition and may be cancelled if there is a change of government.

• See cells A81 : I81 in the cost risk modelling sheet of the example model (Table 3.12).

Strategic risks are the future situations that may utterly confound the objectives of the organisations involved. They may wipe out the investment in terms of money, time, effort and commitment that has been made. A common clue that a risk is strategic is that no money can be rationally allocated to it, as in the above example. The unwanted outcome of a strategic risk is usually, colloquially put, an all bets off situation, or very close to it. If the unwanted outcome occurs the project may continue but not without radical alteration in some regard.

The proper course of action on identifying a risk like the one in the example is to report it so that it can be managed, because it clearly needs to be, and to declare the model as not modelled.

Table 3.12   Example of strategic risk

Row/column	D	E	F	G	H	I
	Synopsis	**Modelling note**	**#1**	**#2**	**#3**	**Sample**
81	The scheme is incompatible with the intentions of Her Majesty's Opposition and may be cancelled if there is a change of government.	Strategic risk. Not modelled.				–

EXAMPLE RISK 11: SAFETY RISK

Safety risk assessments – the prediction of the number of injuries and deaths that will occur in a specific situation over a given number of years has often been the domain of specialists and so does not usually appear in funding risk calculations. Hence their appearance in this section, even though I will still give a example.

I know of only one relavent model that has a safety risk model built into it. I believe the investment analysts who did it were correct to do so because the before-prevention exposure to the financial consequences of the injury and death were considerable. Making the safety risk assessment a component of the financial risk assessment allowed the analysts to put a case to the funders for the provision of safety assurance systems and processes at the time funding was being planned. This avoided a funding hiatus, which would have had to have been introduced retrospectivly, possibly after the funding deals had been closed and possibly after death or injury had happened. The outcome was a tangible, societal benefit; an immediate commitment by the funders to the financing of prevention measures and to the appointment of additional, suitably qualified staff to ensure their effectiveness.

There are books on safety risk assessment[2] that deal with the processes and calculations in depth. The basic approach is very similar to the modelling of the emerging cost chain described in an earlier section. I call it cause consequence loss modelling. I must first point out that what follows should be repeated for all incidents of concern. It works as follows:

An incident of concern is specified. In this case it is the 'Mind the Gap' situation at a railway station when a passenger is confronted with an excessive stepping between the train and the platform (Figure 3.11).

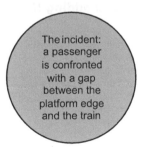

The incident:
a passenger
is confronted
with a gap
between the
platform edge
and the train

Figure 3.11 The incident

The incident is a single person arriving at one gap for one time. What occurs is mapped out as a series of possible events, determined largely by thinking about likely behaviours and through discussions with experts. In this case the consequences are thought to be as follows (Figure 3.12).

---

2   For example Ernest J. Henley and Hiromitsu Kumamoto, Probabalistic risk assessment, published by the Institute of Electrical and Electronic Engineers. ISBN 0-87942-290-4.

The consequences proceed until the incident is over. As shown, there are five possible outcomes.

An accident is an incident in which someone was injured or killed. It is the incident that leads to the accident.

Figure 3.12 shows that no accident will occur in three of the possible five outcomes where the passenger either clears the gap or refuses to attempt to do so.

In Figure 3.13, the losses of each outcome are incorporated.

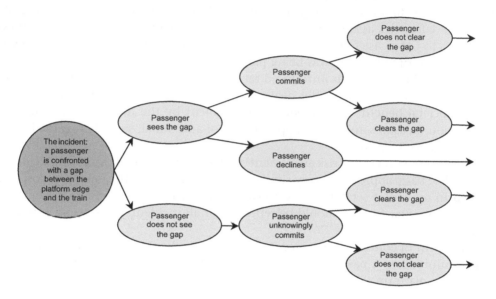

Figure 3.12 Developing the incident: adding the consequences

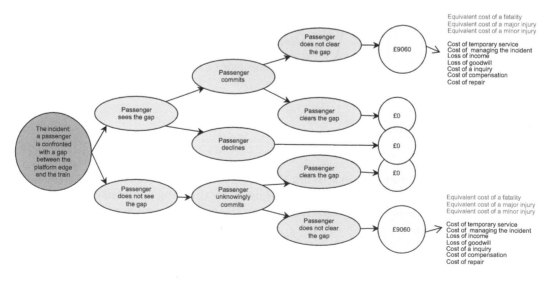

Figure 3.13 Adding in the losses

I have shown a broad a range of potential financial losses:

- Safety-related
  - The equivalent cost of a fatality.
  - The equivalent cost of a major injury.
  - The equivalent cost of minor injury.
- Direct business costs
  - The cost of repairing damage to the assets.
  - The cost or providing temporary facilities and services.
  - The cost of managing the incident.
  - The cost of inquiry and investigation.
  - The cost of compensation for operational disruption.
  - penalties
- Losses
  - Loss of income (a tangible measure).
  - Loss of goodwill (a less tangible measure, probably only assessable by share price).

The equivalent cost of a fatality is a typical cost of a statisitical fatality – not an actual one that may be awarded by the courts as compensation, but a figure derived from an appraisal of the typical losses in terms of income, opportunity and distress that would be caused by a fatality. Its purpose is to help decide, after consideration of the number of fatalities and injuries that could occur, if it would be worthwhile investing in prevention measures. Hence the subtly nuanced title that is formally used among safety professionals – the 'Value of preventing a fatality'. Its value, and those of the subcategories of major injury and minor injury, are determined by the body's responsible for ensuring safety standards are met. In the UK this is the Health and Safety Executive.[3] The actual figure is updated from year to year.

Having modelled the consequences and the losses of the incident, the analyst then has to add in how often the incident happens, Figure 3.14. The number is a function of the number of passengers arriving per unit-time, in this case a year, and so the analyst will need to find out how many trains carrying how many passengers will arrive in that time. A subtlety is shown in the figure: the gap that is to be minded is not just a horizontal one. It could also be a vertical one or a diagonal one. Such nuances are very typical of safety risk assessment work. Analysts preparing safety risk assessments for incorporation into funding submissions should not work in silos but fully discuss their work with the appropriate safety professionals.

The analyst now has a model of how many times per year the incident arises, and when it does, what the consequences may be. Also, if these consequences are undesirable,

---

3   The current Value of Preventing a Fatality in the rail industry is £1.674M and can be found from the Rail Safety and Standards Board website

http://rssb.co.uk/safety/pages/deafult.aspx or
http://docs.google.com/viewer?a=v&q=cache:aWuCKaeOKQYJ:www.rssb.co.uk/SiteCollectionDocuments/pdf/reports/VPF_2010.pdf+cost+of+preventing+an+equivalent+fatality&hl=en&gl=uk&pid=bl&srcid=ADGEEShcSb6HrTypZmhYL7DPlqWtGgcZSztPY8uZ3oPvahts8kbcj1qB1dBb9xye8OEVCJQx_0Xooh11ff4fAB0HUnbv-QD4v3YG91lvrRU5o0DucUu9wvrgZ42oSsWNhaHTLH-xcez9&sig=AHIEtbTM8CtYUKmxioyWK4rSXdbQV5V7Yw.
Information on current evaluations for non-rail scenarios can be obtained from the Department for Transport.
http://www.hse.gov.uk/economics/eauappraisal.htm#refs;
http://www.dft.gov.uk/webtag/documents/expert/pdf/unit3.4.1.pdf.

what financial losses may then arise? An overview of the completed model made up from Figures 3.11, 3.13 and 3.14 is shown in Figure 3.15.

The next step is to quantify the model by incorporating the number of incidents that may occur, the balance of probabilities at each branch in the consequences, and the losses. The expected monetary value (EMV) of the incident for the period of time chosen can then be computed from the summation of products of each loss, and the chain of probabilities that lead to it, multiplied by the number of causes. This will give the overall expected monetary value for the incident. An example is set out component by component in the following set of tables (Tables 3.13, 3.14, 3.15).

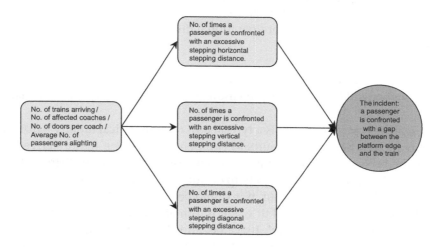

Figure 3.14 The causes – the number of incidents happening over a fixed period of time

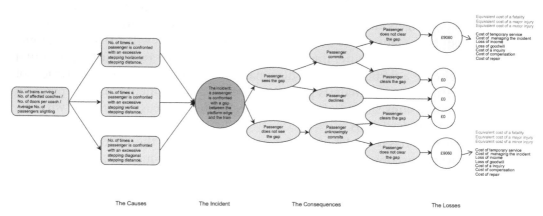

Figure 3.15 The complete cause–consequence–loss model

Table 3.13   Computing the causes

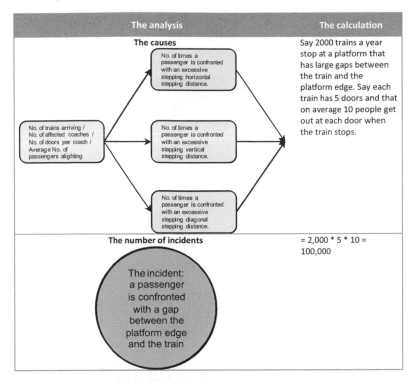

Table 3.14   Computing the consequences

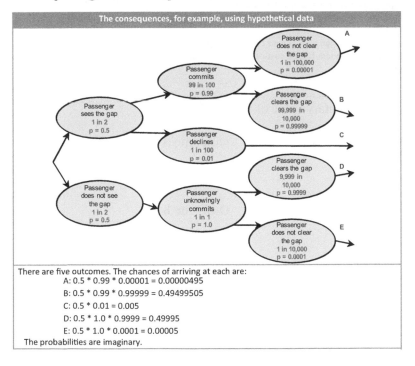

Table 3.15 Computing the losses

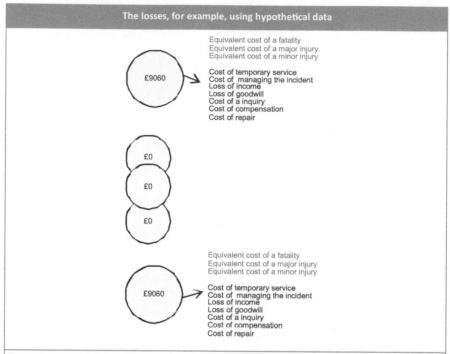

The losses, for example, using hypothetical data

Two outcomes result in accidents – when the passenger does not clear the gap. Say that ...
- 1 in 10,000 accidents incurs a fatality, equivalent cost £1.5M.
- 99 in 10,000 accidents incur a major injury at an equivalent cost of 1/10 of £1.5M.
- 9,900 in 10,000 accidents incur a mi nor injury at an equivalent cost of 1/200 of £1.5M.

The expected loss of each accident would be:

= (0.0001 * £1.5M) + (0.0099 * £1.5M/10) + (0.99 * £1.5M/200)

= £9,060

The expected loss when a passenger clears the gap is £0.

The £1.5M is a rounded, typical cost of a statisitical fatality. It is not the currently **endorsed value of preventing a fatality**.

The expected monetary value of the incident is

0.00000495 * £9,060	Outcome A
+	
0.49499505 * £0	Outcome B
+	
0.005 * £0	Outcome C
+	
0.49995 * £0	Outcome D

+

0.00005 * £9,060   Outcome E

=

£0.489

The risk exposure, per period of time, in this cases a year and based on the EMV, is

100,000 * £0.489 = £48,900 ≈ £50,000

If the operational life of the asset is 30 years, and if the platform gaps were to remain as they are, the total risk exposure would be

30 * £50,000 = £1.5M (before inflation)

If the cost of rectifying the gaps was £1M, the case for additional funding would have a benefit to cost ratio of £1.5M/£1M = 1.5 and the case for funding would be made.

Potential uncertainties in the values of the probabilities assigned to the branching in the consequences tree, in the numbers of incidents and in the values of the losses does allow this type of model to be recast in risk form in which RMT functions would replace spot values. If this was done then the EMV spot exposure would become an EMV spread exposure with associated levels of confidence, but I will pass over this sophistication in order to explain more easily the two final steps in the process.

First, the analyst may want to demonstrate the benefit of putting in risk mitigation measures – the systems and processes intended to reduce the EMV spot exposure – to the funder. Safety risk professionals sometimes call these the control measures. There are three strategies:

• reduce the losses
• influence the consequences towards lower cost outcomes
• reduce the number of incidents

In the example, the losses are determined by the values of the preventable fatalities and the relative proportions in which they are thought to occur. These are immutable because they are externalities: if people fail to mind the gap and fall, 1 in 100,000 will die as a result (these are speculative figures) and the value of that preventable fatality will be £1.5M (again, a speculative figure). Nothing can be done about it.

Furthermore, the number of incidents is fixed by the train timetable and the number of passengers who choose to travel. These too are immutable externalities.

The only possible strategy for control in this case would be to influence the consequences of failing to mind the gap towards a low-cost outcome, hence the 'Mind the Gap' announcements that are a familiar feature of travel on London Underground. These are clearly intended to influence passengers away from not seeing the gap. There is an expectation that this will alter the balance of probabilities at the point in consequences where the passenger may choose to avoid a path, resulting in a £9,060 loss (see Table 3.14 and Table 3.15).

Second, it is important to remember that a safety risk model will have a time frame for the number of incidents (a year in the set of figures above), whereas a control measure will have an operational life of probably many more, and therefore a consistency needs to be introduced. This is so either the number of incidents is adjusted to be the number that would occur over the lifetime of the control, or alternatively the investment in the control is apportioned so that is an equivalent annual cost.

Of all the risk modelling described in this book, safety risk models require the most thorough understanding of human behaviour, particularly our reactions to emerging and changing situations. They should never be developed nor their results published without consulting the safety risk community. More often than not, they can take a considerable time to research, far in excess of the time available to the analyst developing a financial risk model. As a result, if there is a concern among those with whom the analyst is working that existing measures for safety assurance are inadequate, or that the measures intended to ensure construction of project work and their subsequent operations may also be, then the analyst should include a lump sum for their provision on the assumption that they will have to be provided in the near future. Not to do so would be negligent because the funders have not been advised of a foreseeable cost and others have been placed at risk.

In the example model, a fixed sum of £50,000 for additional safety measures has been included, pending full evaluation of the safety risk (Cells A52 : I52).

Time-dependent Cost Primitive

EXAMPLE RISK 12: TIME-DEPENDENT COST

Even if no cost risks of the types described can be identified as possibly likely to affect the project, there will always be this one – the time-dependent cost, which I will discuss in the next but one section about the time risk part of the model.

Digressions

ON MODELLING RISK TO THE ENVIRONMENT

This is modelled in the same way as safety risk assessment except that instead of incidents that may lead to injury and death, the scope of undesirable outcomes is broader. These include noise levels, loss of diversity of flora and fauna, increases in pollutant levels, loss of utility and loss of amenity. None of these can be quantified in financial terms that would be agreed as rationally based and compellingly correct, unlike the legal basis and established use of earnings lost over an expected lifetime that supports acceptance of the injury and death evaluations.

Faced with this problem of there being no satisfactory, consensual financial measures with which to assess environment risk, I suggest the best option for the analyst is to change the currency of the assessment from pounds and weeks that could be lost to measures that are more appropriate, for example, to size of population or levels of pollutant.

By altering the currencies of the risk calculations, in effect, by re-casting the cause–consequence–loss model used the safety risk assessment so that is a threat to the environment rather than to the individual that is the incident at its core, and the consequential changes in measures of the quality of the environment that becomes its losses, the analyst then has a way of reporting risk to the environment to the funding community. What financial value it then places on the losses is for it and the political community to argue and agree. The analyst can do no more.

On the Computational Framework for a Cost Risk Model

The computational framework for a cost risk model is as you may expect – a columnar spreadsheet summation of the outputs of the individual risk model formulae. The risk descriptions and calculations tend to flow from left to right, culminating in a sample of their evaluation. The samples are then vertically summed (Figure 3.16).

Large models may contain extensive submodels whose outputs are consolidated into a main sheet. The individual risk model components may calculate interim results such

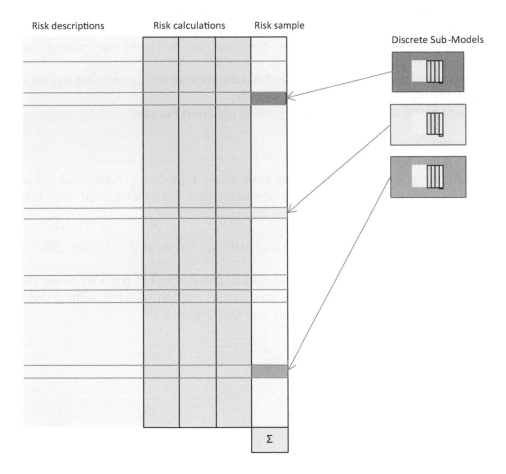

Figure 3.16 A typical computational framework for a cost risk model

that they spread over a wide and deep number of cells, but every risk model I have seen has a right-hand side column into which samples from the formulae are deposited for summation at the top or the bottom.

This is obvious, so why explain it? The reason is that the computational framework for a time risk model is different and not quite as obvious.

On the Computational Framework for a Time Risk Model – the Spreadsheet Emulation

The calculation used to work out how long a project will take is called the forward pass. Here, zoomed out in Figure 3.17 (it is not necessary to read the detailed task descriptors in the boxes in the figure for what follows), is the logic network from Figure 2.9.

A forward pass is the summation of the individual task durations from left to right along the paths on which they are placed. Where two or more paths join, the largest of the summations thus far is carried forward as the start value of the next duration additions. In Figure 3.17 this gives an overall duration of the 4 January 2010 to 18 October 2013, 198 calendar i.e. elapsed, weeks.

The thing to recognise is that the forward pass calculation can be done in Excel. A project planning tool is not needed. Excel provides everything an analyst will need to emulate the project plan and incorporate time risks into it. The forward pass calculation takes the following form at each task:

Finish week number = start week number + duration of task in calendar weeks (where the granularity of time being modelled is weeks).

I suggest weeks is the best choice regarding precision because:

- A month or a year is too coarse
- A day or an hour is too fine
- A calendar week will cover a working week of any type: 5 day, 7 day, weekend only
- To an inaccurate but often practical approximation, 0.1 of a calendar week is half a day.

The forward pass calculation is easy to translate into Excel, as it is a daisy chain set of sequential tasks (Figure 3.18).

These tasks are all linked by finish to start relationships, but it is easy to see that by changing the references in the start cells, start to start relationships can be modelled; with the addition of some Excel arithmetic, so may lags (Figure 3.19).

Figure 3.17 The logic linked network from Figure 2.9

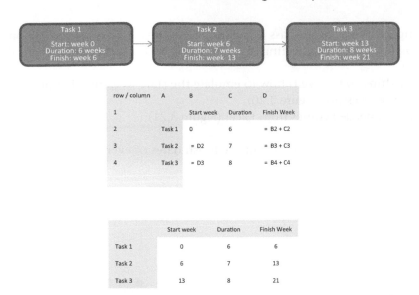

Figure 3.18 A set of sequential tasks emulated in Excel in a forward pass calculation

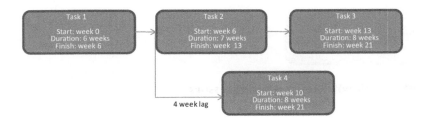

Figure 3.19 A set of sequential tasks emulated in Excel in a forward pass calculation, with a start to start relationship and a lag

With ingenuity-backward pass calculations that reveal how much freedom there is, rescheduling tasks without affecting the completion date and all types of logical links can be emulated.

The only thing to resolve is how to emulate the start week number of a task that has two or more predecessors (Figure 3.20).

The formula needed comes from Excel (Figure 3.21):

Start week of task = MAX(a list of the finish dates of every preceding task)

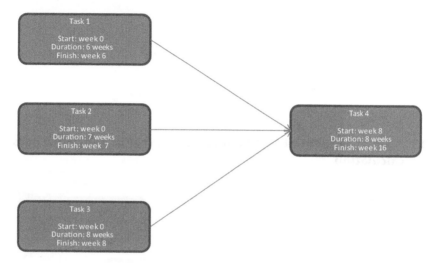

Figure 3.20 A task with several predecessors

row / column	A	B	C	D
1		Start week	Duration	Finish Week
2	Task 1	0	6	= B2 + C2
3	Task 2	0	7	= B3 + C3
4	Task 3	0	8	= B4 + C4
5	Task 4	= MAX(D2:D4)	8	= B5 + C5

	Start week	Duration	Finish Week
Task 1	0	6	6
Task 2	0	7	7
Task 3	0	8	8
Task 4	8	8	16

Figure 3.21 An emulation of a task with several predecessors

The previous three figures showed emulation in Excel applied to a plan of just three of four tasks, but it's easy to see that it can be applied to a plan containing hundreds. All of the durations can usually be extracted by cut and paste from the plan as it exists in the planning software, perhaps Microsoft Project or Primavera P3. All of the finish week numbers use the same arithmetic formula: end = start + duration. The only tricky task is to emulate the logic of the plan by correctly referencing each task's start to its predecessor or predecessors. There are two quick ways to do this.

One – most tasks will have finish-to-start relationships in which the start week of a task is the finish week of its predecessor. This means the emulation is a mostly a single chain of tasks and its emulation an easy copy-down-the-column spreadsheet operation of one cell, the start day, equalling another, the finish day. Then specifically edit the start weeks that are not sequential but instead are concurrent. For example, in Figure 3.22, every task follows directly on from its numerical predecessor: #2 follows #1, #3 follows #2 and so on, except for the four encircled: #5 does not follow #4 but instead #1, #8 follows #5, #10 follows #4 and #7, and #11 follows #9 and #10. Only these four tasks need to have specific start-week logic encoded.

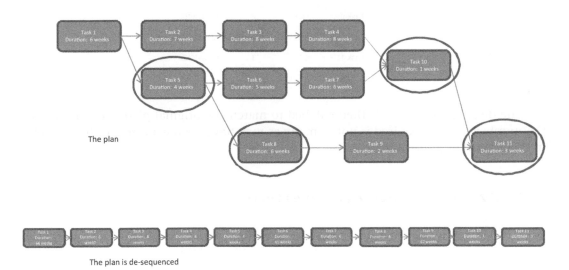

Figure 3.22 De-sequencing the logic of a plan

The de-sequenced logic is encoded (Table 3.16):

Table 3.16   The de-sequenced logic of the plan encoded

		J	K	L
		Start week	Duration	Finish Week
2	Task 1	0	6	=J2+K2
3	Task 2	=L2	7	=J3+K3
4	Task 3	=L3	8	=J4+K4
5	Task 4	=L4	8	=J5+K5
6	Task 5	=L5	4	=J6+K6
7	Task 6	=L6	5	=J7+K7
8	Task 7	=L7	6	=J8+K8
9	Task 8	=L8	8	=J9+K9
10	Task 9	=L9	2	=J10+K10
11	Task 10	=L10	1	=J11+K11
12	Task 11	=L11	3	=J12+K12

The de-sequenced logic is then patched to match the original plan. As you can see, the number of edits required is not as many as may have been expected; only four cells in this typical example (Table 3.17):

Table 3.17   The sequenced logic of the plan encoded

		J	K	L
		Start week	Duration	Finish Week
2	Task 1	0	6	=J2+K2
3	Task 2	=L2	7	=J3+K3
4	Task 3	=L3	8	=J4+K4
5	Task 4	=L4	8	=J5+K5
6	Task 5	=L2	4	=J6+K6
7	Task 6	=L6	5	=J7+K7
8	Task 7	=L7	6	=J8+K8
9	Task 8	=L6	8	=J9+K9
10	Task 9	=L9	2	=J10+K10
11	Task 10	=MAX(L5,L8)	1	=J11+K11
12	Task 11	=MAX(L10,L11)	3	=J12+K12

Two – planning software invariably gives each task a number. It is quite an easy typographical error when encoding emulations to confuse the task number with the spreadsheet row number. In the example above, Task 6 appears in row 7. Being a successor to Task 5, its start needs to refer to row 6 but it is all too easy to encode a reference to row 5. Tracing such bugs can be very time-consuming for large emulations and so there is a quick way to ensure the row number on which a task appears equals its number + 10, thus (Table 3.18):

Table 3.18 Matching task number to (row number + 10)

		J	K	L
		Start week	Duration	Finish week
11	Task 1	0	6	= J11 + K11
12	Task 2	= L11	7	= J12 + K12
13	Task 3	= L12	8	= J13 + K13
14	Task 4	= L13	8	= J14 + K14
15	Task 5	= L11	4	= J15 + K15
16	Task 6	= L15	5	= J16 + K16
17	Task 7	= L16	6	= J17 + K17
18	Task 8	= L15	8	= J18 + K18
19	Task 9	= L18	2	= J19 + K19
20	Task 10	= MAX(L14,L17)	1	= J20 + K20
21	Task 11	= MAX(L19,L20)	3	= J21 + K21

Task 1 is in spreadsheet row 11, and so on. Thus, when working deep inside an emulation if I need to make Task 657 start when task 345 has finished, it would be a reference to Row 345 + 10, Row 355. If task 78 starts when tasks 23, 156 and 289 have finished then my formula for the start would be = max (L33, L166, L299), assuming the end dates appeared in column L.

These two shortcuts make emulations very quick to encode. I would not baulk at emulating a typical 1000-activity programme within a working day, and in practice, when asked to assess a project for which a plan does not yet exist and so for which I need to create one, I usually start creating the emulation from scratch and give the completed logic back to the planning team so they can incorporate it into their software tools. They may well add in resource constraints and a range of calendars reflecting different types of working weeks that introduce divergences between a project end date, and the emulation. I may then have to model these complications in the emulation, but nine times out of ten I do not because the plans at the stage of the development cycle that most risk assessments are carried out are simply not that far advanced. Further, I do not need to know the precise start and finish dates of each activity because I am not filling in the project diary. I only need to know by how much the project may overrun. All planners will ruefully acknowledge that they are never asked what this will be, but are told it anyway. If my emulation in its unrisked form matches the given end date, then the plan is almost sure to compute the same date too.

The final step in the emulation is to add risk to it. Here the computational framework for time risk differs from cost risk. It is still spreadsheet trickery but whereas cost risk models tend to model one risk per row, time risk models are erratically disordered as opposed to neatly monolithic. The basic idea is to take a copy of an emulation and to replace the values in the task duration cells with risked-up versions of that duration, thus:

End week = Start week + 3 weeks.

May become

End week = Start week + RMT/Uniform (2,4) weeks.

The erratic nature of the time risk model manifests itself when more than one risk affects a single duration, and so the emulation needs to be referenced to a subset of risk-modelling computations whose combined effect is copied back into the emulation (Figure 3.23). This inevitably does not make for the neat, tidy look to our work that many of us try to sustain.

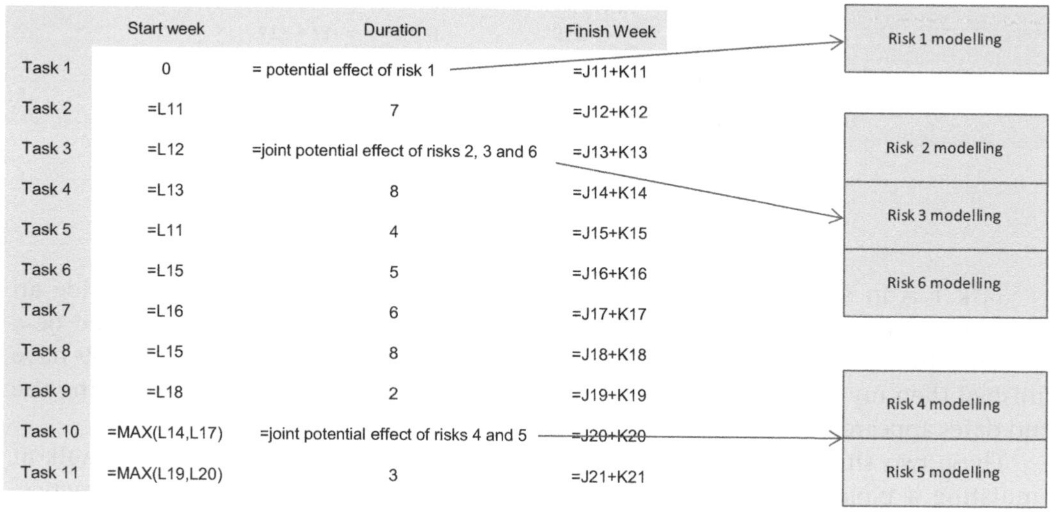

	Start week	Duration	Finish Week	
Task 1	0	= potential effect of risk 1	=J11+K11	Risk 1 modelling
Task 2	=L11	7	=J12+K12	
Task 3	=L12	=joint potential effect of risks 2, 3 and 6	=J13+K13	Risk 2 modelling
Task 4	=L13	8	=J14+K14	Risk 3 modelling
Task 5	=L11	4	=J15+K15	
Task 6	=L15	5	=J16+K16	Risk 6 modelling
Task 7	=L16	6	=J17+K17	
Task 8	=L15	8	=J18+K18	
Task 9	=L18	2	=J19+K19	Risk 4 modelling
Task 10	=MAX(L14,L17)	=joint potential effect of risks 4 and 5	=J20+K20	
Task 11	=MAX(L19,L20)	3	=J21+K21	Risk 5 modelling

Figure 3.23 The erratic nature of an emulation

We can now emulate the logic, and compute the project completion date, because all values in the emulation of a logic linked plan are in Excel. They are available to be replaced or modified by risk formulae that reflect the behaviour of the time risks.

The path with the longest total duration through the network determines the overall duration of the project (it cannot be done any quicker according to the plan) and is the critical path. There may be more than one critical path if there are equal longest duration paths.

Such a computational framework for a time risk model is a very powerful tool. I know of emulations that include multiple calendars and resource constraints.

To calculate actual dates for the starts and finishes of tasks in an emulation requires just a simple Excel addition of the lapsed weeks to a reference date. An example is shown in Figure 3.24.

row / column	A	B	C	D	E	F	G	H
1								
2	Task No.	Task	start	predecessor task id	duration	finish	start date	finish date
3	1	prepare Business Case paper for authorisation for Feasibility and Outline Design	0	-	4	= C3 + E3	= DATE(2010,3,1)	= $G$3 + F3*7
4	2	obtain Business Case authorisation	= F3	1	1	= C4 + E4	= $G$3 + C4*7	= $G$3 + F4*7

Figure 3.24 The calculation of actual dates in an emulation

Two tasks are shown encoded in rows 2 and 3 of the emulation. Their start week numbers are determined in column C, and their finish week numbers in column F. These are converted in columns G and H to actual calendar dates relative to a fixed-start week encoded in cell G3. This was achieved by the simple addition of the number of weeks that have lapsed since the fixed-start date converted to days by the $7 \times$ multiplication.

4 *Risk Modelling Examples: Time*

An Overview of the Example Model (Time Part, the Emulation)

Here is the full emulation (Figure 4.1). A somewhat abbreviated list of tasks for a construction project is set out in column B, from Preparation of the business case to Packing up the site. Each Task has a Task number (column A) and the logical linking of the tasks in terms of those numbers is set out in column D – Predecessors. The logical linking in terms of spreadsheet row numbers and column letters is emulated in column C – Start. The emulation is completed by inclusion of each task's duration (column E) and calculation of its finish week (column F). The actual dates that each task starts and finishes are calculated in columns G and H relative to a fixed start date in cell G3. The overall 'spot' completion week and date is determined in cell F25. The 'spot' value is a colloquialism used to distinguish a single valued result from its equivalent risk-based one, the 'spread' value.

Some cells are left blank because the rows in which they are situated will be used for risk formulae.

Figure 4.2 shows a zoomed-out image of the emulation alongside its risk model equivalent. As on earlier occasions, it is not necessary to read the detail in order to understand what follows.

The light grey cells in the spot calculation to the left of a heavy black line are re-coloured a darker grey in the copy on the right, which is the spread calculation. Lighter grey areas within the darker grey contain an analysis of the risks that may affect to the task entered in the row, and the modelling of the possible impact of these on that task's duration. The emulation and the calculation of the finish dates remain unchanged. Only the durations are different because of the effect of the risks. Speckled columns contain the calculations that convert lapsed weeks to calendar dates.

Spread Duration Primitive

EXAMPLE RISK 13: CONTINUOUS SPREAD DURATION

Spread duration risks most often occur as productivity variations due, for example, to labour output or process overload.

Several of the tasks in the emulation have spread durations, for example, Task 3: Appoint consultants, looks like this in the spot calculations.

- See A5 : V5 in the example time risk model sheet (Figure 4.2).

row / column	A	B	C	D	E	F	G	H
1								
2	Task No.	Task	start	predecessor task numbers	duration	finish	start date	finish date
3	1	prepare Business Case paper for authorisation for Feasibility and Outline Design	0	-	4	= C3 + E3	= DATE (2010,3,1)	= $G$3 + F3 * 7
4	2	obtain Business Case authorisation	= F3	1	1	= C4 + E4	= $G$3 + C4 * 7	= $G$3 + F4 * 7
5	3	appoint consultants	= F4	2	4	= C5 + E5	= $G$3 + C5 * 7	= $G$3 + F5 * 7
6	4	carry out Feasibility and Outline Design	= F5	3	24	= C6 + E6	= $G$3 + C6 * 7	= $G$3 + F6 * 7
7	5	Sponsors assess Feasibility Study	= C6+4	4SS+4w	8	= C7 + E7	= $G$3 + C7 * 7	= $G$3 + F7 * 7
8	6	prepare Business Case paper for authorisation for Detailed Design and Implementation	= MAX (C6+10,F7)	4SS+10, 5	4	= C8 + E8	= $G$3 + C8 * 7	= $G$3 + F8 * 7
9	7	obtain Business Case authority	= F8	6	4	= C9 + E9	= $G$3 + C9 * 7	= $G$3 + F9 * 7
10	8	tender for services of consultants and contractors	= F9	7	16	= C10 + E10	= $G$3 + C10 * 7	= $G$3 + F10 * 7
11	9	appoint consultants and contractors	= F10	8	8	= C11 + E11	= $G$3 + C11 * 7	= $G$3 + F11 * 7
12	10	carry out surveys and do detailed design: all disciplines	= F11	9	24	= C12 + E12	= $G$3 + C12 * 7	= $G$3 + F12 * 7
13	11	clear work site of existing infrastructure	= F11	9	3	= C13 + E13	= $G$3 + C13 * 7	= $G$3 + F13 * 7
14	12	consult affected parties	= F6	4	24	= C14 + E14	= $G$3 + C14 * 7	= $G$3 + F14 * 7
15	13	mobilise the piling contractor and construct piles	= C13+4	11SS+4	12	= C15 + E15	= $G$3 + C15 * 7	= $G$3 + F15 * 7

row / column	A	B	C	D	E	F	G	H
16	14	mobilise the other contractors	= MAX (F12,F13,F14,F15)	10, 11, 12, 13	6	= C16 + E16	= $G$3 + C16 * 7	= $G$3 + F16 * 7
17	15	erect signals	= F16		5	= C17 + E17	= $G$3 + C17 * 7	= $G$3 + F17 * 7
18	16	do general civils	= F16	14	24	= C18 + E18	= $G$3 + C18 * 7	= $G$3 + F18 * 7
19	17	< left blank >						
20	-	< left blank >						
21	18	do rail track civils	= C18+8	16SS+8	12	= C21 + E21	= $G$3 + C21 * 7	= $G$3 + F21 * 7
22	19	do signalling installation mods	= F16	14	12	= C22 + E22	= $G$3 + C22 * 7	= $G$3 + F22 * 7
23	20	commission systems	= MAX (F17,F18,F21,F22)	15, 16, 18, 19	4	= C23 + E23	= $G$3 + C23 * 7	= $G$3 + F23 * 7
24	21	re-instate overhead wiring	= F23	20	2	= C24 + E24	= $G$3 + C24 * 7	= $G$3 + F24 * 7
25	.	< left blank >						
26	.	< left blank >						
27	.	< left blank >						
28	.	< left blank >						
29	.	< left blank >						
30	.	< left blank >						
31	.	< left blank >						
32	.	< left blank >						
33	22	evaluation period	= F24	21	4	= C46 + E46	= $G$3 + C46 * 7	= $G$3 + F46 * 7
34	23	pack up site	= F46	21	4	= C47 + E47	= $G$3 + C47 * 7	= $G$3 + F47 * 7
35	24	landscape the site	= F47	23	0	= C48 + E48	= $G$3 + C48 * 7	= $G$3 + F48 * 7
36	25							
37	26							
38					spot >	= F47		

Figure 4.1   A Project Plan Emulation

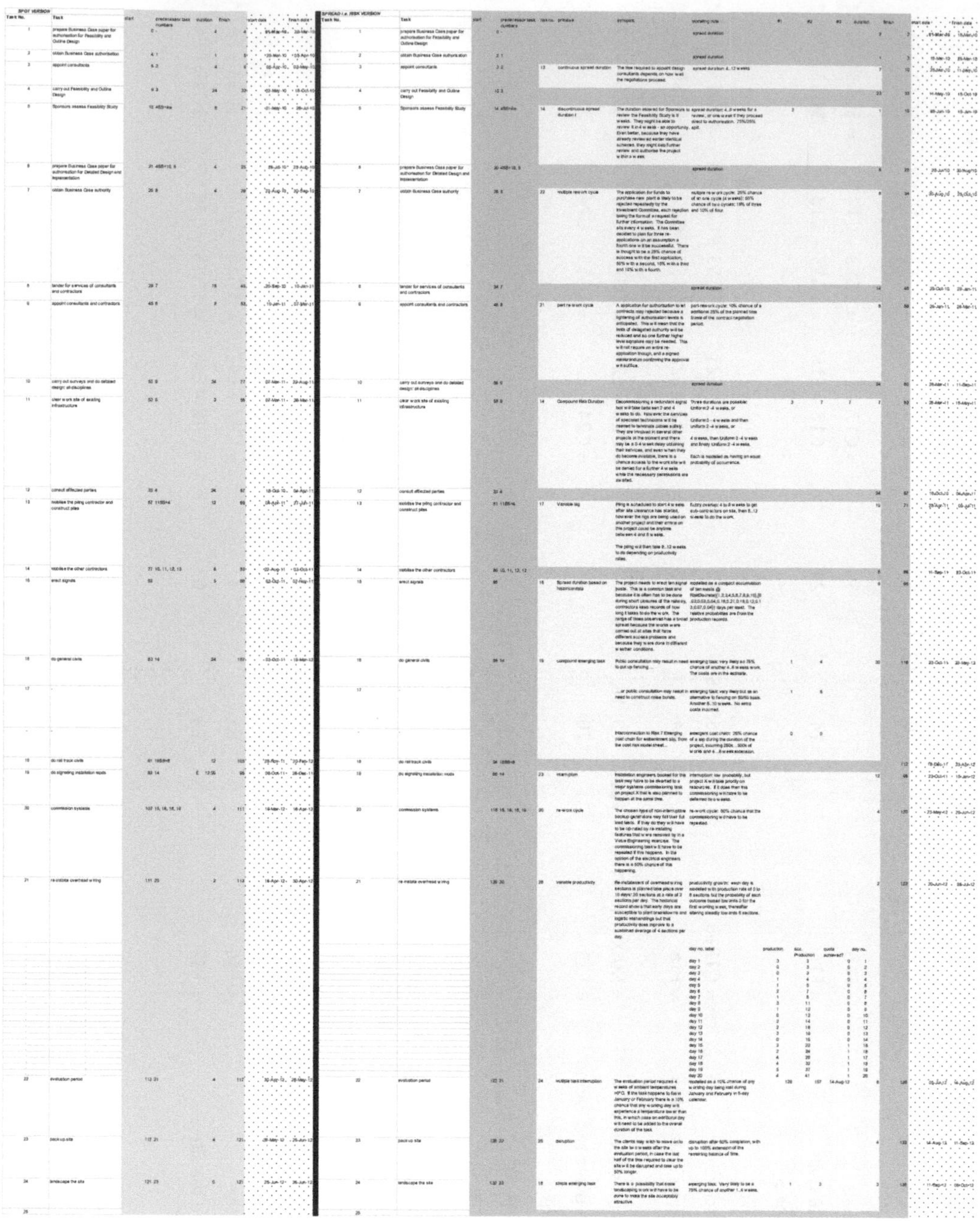

Figure 4.2   The spot emulation and the spread emulation (zoomed-out overview)

It becomes, in its spread calculation neighbour, Figure 4.3.

Task No.	Task	start	predecessor task numbers	duration	finish
3	Appoint **consultants**	= F4	2	4	= C5 + E5

Figure 4.3   A Spot Duration

Note how the simple duration in the former has been broadened by the inclusion of columns for risk synopsis, modelling note, interim calculations #1 and #2 and by a risk formulation, instead of a number in the duration cell itself.

The risk modelling recognises the reality that the four-week duration for the appointment in the plan is an aspiration. The risk analyst has (presumably) sounded out other opinions and decided that 4–12 weeks is a more likely time frame for consultants to be available in sufficient numbers to commence the succeeding task in the plan, Carry out feasibility study.

Notice the subtlety – the analyst is not saying the manager is wrong. If the manager wishes to impose a target on themself of four weeks then that it is their business. But the analyst represents the interests of the sponsors and the funders, and here, 4–12 weeks has been (presumably) found to be the most realistic outcome.

In the emulation there is now a possibility of the task taking between 4–12 weeks to complete. On successive iterations of the risk model, a new random value between 4 and 12 will appear in the task duration cell. That new value will be counted in the forward pass calculation of the emulation in the same iteration, and one of three things will happen.

- If the task was on the critical path already with a four-week duration, the duration of the project will increase.
- If the task was not on the critical path, an increase in duration to more than four weeks as a consequence of an iteration may put it on the critical path, in which case the duration of the project will increase.
- The task may not be on the critical path even when the duration is eight weeks, in which case the risk does not affect the duration of the project.

Task No.	Task	start	predecessor task numbers	synopsis	modelling note	#1	#2	duration	finish
3	appoint consultants	= Q4	2	The time required to appoint design consultants depends on how well the negotiations proceed.	*spread duration*			= RMT/ Uniform (4,12)	= J5 + P5

Figure 4.4   The Spread Calculation Equivalent

If, in a modelling run, we were to monitor the overall project duration cell in the emulation, we could produce a set of statistics for the time risk exposure equivalent to that generated for the cost risk exposure.

Notice also that the time frame, 4 to 12 weeks is a continuum. This is not always the case, as shown in the next example.

EXAMPLE RISK 14: DISCONTINUOUS SPREAD DURATION

Task 5: Sponsors assess feasibility study has a duration of eight weeks in the spot calculation but is subject to the following risk:

The duration allowed for sponsors to review the feasibility study is eight weeks. They might be able to review it in four weeks – an opportunity. Even better, having reviewed earlier identical schemes they might skip further review and authorise within a week.

The model is illustrated in Table 4.1. To facilitate side-by-side comparison only the description and quantification columns are shown in the Table 4.1. The emulation columns are hidden.

Table 4.1   Example of a discontinuous spread duration: 1

Row/ column	A	B	E	P	Q	R	U
	Task no.	Task	Spot duration	Risk	Modelling note	#1	Spread duration
7	5	Sponsors assess feasibility study	8	The duration allowed for sponsors to review the feasibility study is 8 weeks. They might be able to review it in 4 weeks – an opportunity. Even better, because they have already reviewed earlier identical schemes, they might skip further review and authorise the project within a week.	Spread duration: 4–8 weeks for a review, or one week if they proceed direct to authorisation. 75%/25% split.	= RMT/ Discrete ({1,2}, {75,25})	= IF(N7 = 1, RMT/Uniform (4,8), 1)

See Cells A7 to V7 in the time risk modelling sheet of the example model.

Here the analyst has found that the sponsors sometimes take less time than the planner has pessimistically presumed and has modelled this as an opportunity – the duration will be either four to eight weeks, not eight, or perhaps even as low as one week if the study is nodded through without further ado.

Which of these two outcomes pertains is modelled in Interim Calculation #1, with the consequent durations in the adjacent spread duration cell?

Similarly to the previous example, the continuous spread duration, there is now notionally an opportunity for this task to take either four to eight weeks or just one week. On successive iterations of the risk model, a new random value between four and eight (or one) will appear in the task duration cell. That new value will be counted in the forward pass calculation of the emulation in the same iteration. One of two things will happen:

- If the task was on the critical path with an eight-week duration, the duration of the project will decrease if the opportunity evaluates to less than eight weeks on an iteration.
- If the task was not on the critical path even when the duration was eight weeks, the opportunity gained will not affect the duration of the project.

A subtlety arises from the last point: if a task is not on a critical path, is it worthwhile pursuing an opportunity to make it shorter? If the costs of the task are time-dependent in some way then perhaps yes, but if they are not, then perhaps no, because the overall duration of the project will not be altered, though the amount of slack might increase and more time flexibility may be provided.

Indeed, having one task finish early can be a bit of a nuisance because its resources will then need to be redeployed or decommissioned prematurely, possibly disruptively and certainly out of sequence Therefore, though all risks threaten a project, not all opportunities necessarily benefit it. They can be disruptive.

EXAMPLE RISK 15: COMPOUND SPREAD DURATION

Here is an example of a slightly different discontinuous distribution I have devised to illustrate a flaw in the wide use of triple-point estimates for almost everything in risk modelling.

- Decommissioning a redundant signal box will take between two and four weeks to do. However, the services of specialist technicians will be needed to terminate cables safely. They are involved in several other projects at the moment and there may be a 0–4-week delay obtaining their services, and even when they do become available, there is a chance access to the work site will be denied for a further four weeks while the necessary permissions are awaited.
- See cells A13 to V13 in the time risk modelling sheet of the example model.

The model looks as it does in Table 4.2. To facilitate side-by-side comparison only the description and quantification columns are shown in Table 4.2. The other columns are not shown.

Table 4.2   Example of a discontinuous spread duration: 2

Row/column	A	B	E	Q
	Task no. Task		Spot duration	Spread duration Modelling note
13	11	Clear work site of existing infrastructure	3	Three durations are possible: Uniform 2–4 weeks, or Uniform 0–4 weeks and then uniform 2–4 weeks, or 4 weeks, then Uniform 0–4 weeks and finally Uniform 2–4 weeks. Each is modelled as having an equal probability of occurrence.

Three durations are possible:

1. Uniform 2–4 weeks to do the work if there are no hold ups at it start, or
2. Uniform 0–4 weeks and then Uniform 2–4 weeks if there is a delay waiting for the specialists to mobilise from another job, or
3. The same situation but with an additional four-week delay if the access the site is held over by (presumably) a planning authority of some sort.

Each is modelled in cells R13, S13 and T13 with an equal probability of occurrence set in cell U13 by an Excel Choose function in which a RMT/Discrete ({1,2,3}, {1,1,1}) is embedded.

The spread for this risk is shown in Figure 4.5.

However, the risk could have been naively modelled by the risk analyst as:

= RMT/Triangular (Minimum, Most Likely, Maximum)

where

1. the Minimum = 2, from the minimum of 2–4 duration when there is no hold-up to starting the work
2. the Maximum = 12, when there is a four-week holdover on permission, followed by a four-week mobilisation delay and then four weeks of work
3. and the Most Likely chosen to be seven, an arbitrary choice made because it is halfway between the minimum and the maximum.

= RMT/Triang (2, 7, 12)

The duration spread of this modelling is shown in Figure 4.6 on top of the correct modelling.

The difference is clear. The compounded duration has a completely different profile, with a much more pronounced tendency to generate generations in the region of two to four weeks than the triangle function. However, since both the correct and incorrect models produce answers of the same order of magnitude with the same maximum,

12, why bother with the former? Is it so very wrong to generate 2–12 from a triangular function as it is to generate 2–12 from a compounding of uniform functions? Does it matter? My answer to this is a deeply held conviction that yes, it does. Risk modelling does not have a good press and many of my clients are extremely sceptical about it. I think one contributor to this regrettable situation is the prevalence of unrealistic and naive modelling that is inconsistent with the experience and expectation of those for whom the modelling is being done. The predominant, in some cases exclusive, use of minimum,

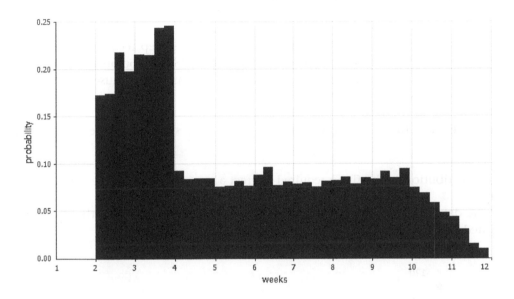

Figure 4.5   The duration spread for the compound spread duration example

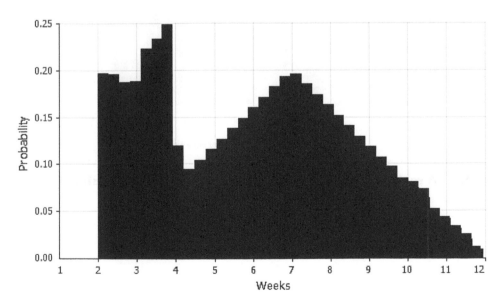

Figure 4.6   Correct and incorrect modelling

most likely and maximum parameters for every risk, when a moment's consideration should raise doubts that cannot be right, has to be a contributor to this general lack of credibility. If I cannot be bothered to model the risks correctly, why then should a client be bothered to trust my results?

Besides, how much longer does it take to encode cells R13 to U13 than it does RMT/ Triang (2,7,12)? A few tens of seconds, that is all.

EXAMPLE RISK 16: SPREAD DURATION BASED ON HISTORICAL DATA

The project needs to erect ten signal posts. This is a common task and because it is often done during short closures of the railway, the contractors keep records of how long it takes to do the work. The range of times noted has a broad spread because the works have been carried out at sites that have different access problems, and because they have been done in different weather conditions.

- See cells A17 to V17 in the time risk modelling sheet of the example model (Figure 4.7).

The distribution of the times required to erect a post was found to be between 1 and 10 days.

If the data had, for some inexplicable reason, been found to have the distribution illustrated in Figure 4.8:

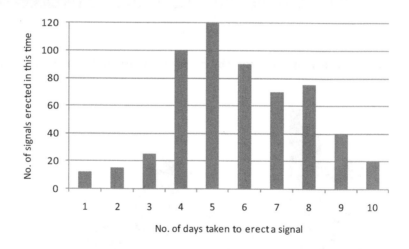

Figure 4.7   A realistic distribution of the times taken to erect a signal

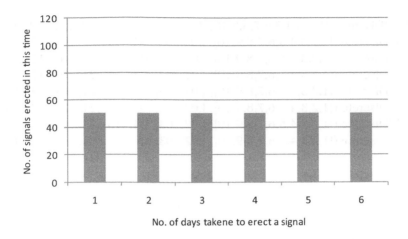

Figure 4.8   An unrealistic distribution of times taken to erect a signal

we would have little difficulty deciding this uncertainty should be modelled by a 1–6 dice. The number on the dice would give an indication to the number of days taken, and each outcome would have an equal probability of occurring. The actual uncertainty shown in the penultimate figure, Figure 4.7, is simply a 1–10 dice weighted so that the values have different probabilities of occurrence. We have seen something similar in the previous sections on cost modelling. The model of the time it could take to erect a signal is simply an RMT discrete function based on the relative probabilities of the different durations:

Days	No.	Probability = No./total
1	12	0.02
2	15	0.03
3	25	0.04
4	100	0.18
5	120	0.21
6	90	0.16
7	70	0.12
8	75	0.13
9	40	0.07
10	20	0.04
	567	1.00

= RMT/Discrete ({1,2,3,4,5,6,7,8,9,10}, {.02,.03, .04, .18, .21, .16, .12, .13, .07, .04})

However, that is only for a single signal. To model the time taken for ten (ten times the above) would be wrong because of the implicit, false, assumption that each erection will take the same time. The formula needs to be instantiated ten times and the total divided seven to give the overall duration in weeks…

= (RMT/Discrete ({1,2,3,4,5,6,7,8,9,10}, {.02, .03, .04, .18, .21, .16, .12, .13, .07, .04})
+ RMT/Discrete ({1,2,3,4,5,6,7,8,9,10}, {.02, .03, .04, .18, .21, .16, .12, .13, .07, .04})
+ RMT/Discrete ({1,2,3,4,5,6,7,8,9,10}, {.02, .03, .04, .18, .21, .16, .12, .13, .07, .04})
+ RMT/Discrete ({1,2,3,4,5,6,7,8,9,10}, {.02, .03, .04, .18, .21, .16, .12, .13, .07, .04})
+ RMT/Discrete ({1,2,3,4,5,6,7,8,9,10}, {.02, .03, .04, .18, .21, .16, .12, .13, .07, .04})
+ RMT/Discrete ({1,2,3,4,5,6,7,8,9,10}, {.02, .03, .04, .18, .21, .16, .12, .13, .07, .04})
+ RMT/Discrete ({1,2,3,4,5,6,7,8,9,10}, {.02, .03, .04, .18, .21, .16, .12, .13, .07, .04})
+ RMT/Discrete ({1,2,3,4,5,6,7,8,9,10}, {.02, .03, .04, .18, .21, .16, .12, .13, .07, .04})
+ RMT/Discrete ({1,2,3,4,5,6,7,8,9,10}, {.02, .03, .04, .18, .21, .16, .12, .13, .07, .04}) +
RMT/Discrete ({1,2,3,4,5,6,7,8,9,10}, {.02, .03, .04, .18, .21, .16, .12, .13, .07, .04}))/7

which though its size and clutter may offend those who prefer to see conciseness and simplicity, is quick and easy to encode (one write, one copy, nine pastes – with minor twiddles and tweaks for the arithmetic operators and brackets) and moreover, is the right thing to do.

There are two more ways this bespoke curve can be derived. These may be practical because they operate directly on the raw data, without the need to compute the interim result of a probability density function to find out the parameters required for the RMT Discrete function. I find a density function a bit fiddly to compute in Excel because it requires the use of the Histogram tool (it is within the Data Analysis add-in tool pack, which if it is installed, is accessible through the Analysis menu on the Data ribbon).

RMTs usually provide two ways to achieve this. First, the advanced versions often provide a distribution fitting tool, which will generate the closest RMT function to the data, effectively taking away from the analyst the interim task of generating a histogram. This a very powerful tool that nicely reveals why advanced RMTs have many more risk functions than one may think will be needed for risk modelling. The closest matching shape to the data can sometimes be an obscure one.

In Figure 4.9 I have used the tool to fit a curve to 1000 records of how many days it took to erect ten signals.

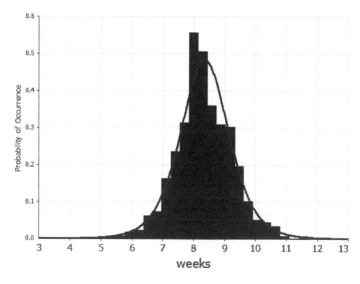

Figure 4.9   A curve fitted to a range of data

The histogram of the durations is shown as a set of blue columns and the best-fitting curve the tool can find is a red line that envelops them. The RMT distribution that best fits the results is one from the less visited corners of an RMT distribution library:

= RMT/Loglogistic (-129.68,138.03,266.55)

The result says that the ten discrete functions used in an ensemble to model the risk could be replaced by a single Loglogistic function, and that this would be a more accurate way to model the risk than would a Min/Most Likely/Max Triangular function, if only the analyst could only elicit the –129.68, 138.03 and 266.55 parameters from the maintenance men. It is not necessary to do this, however, because the only parameters needed are the range of times it takes to erect one signal.

As may be anticipated, the fitting tool does require experience and judgement to use it confidently, even more so than Excel's histogram, and so by far the easiest solution is to compute the cumulative percentiles of the data and to place these in a RMT cumulative function. Here is the computation done at 5 per cent intervals:

- Take 500 data samples (say) taken from the maintenance records. Each is the number of days taken to erect one signal:

  1, 2, 3, 8, 8, 7, 4, 7, 8, 8, 2, 7, 6, 10, 4, 4, 5, 5, 6, 5, 6, 7, 7, 7, 5, 6, 4, 6, 5, 7, 4, 10, 8, 6, 7 etc.

- Calculate the percentile values at 5 per cent intervals, including the minimum and maximum of the data set. The cell references are to the data and the percentile parameter (Table 4.3):

Table 4.3   Example of an equivalent function using cumulative probabilities

The percentile	The Excel formula	The percentile value
Minimum	= MIN($D$3:$D$502)	5.57
5	= PERCENTILE($D$3:$D$502,$F4)	6.71
10.0	= PERCENTILE($D$3:$D$502,$F5)	7.14
15.0	= PERCENTILE($D$3:$D$502,$F6)	7.43
20.0	= PERCENTILE($D$3:$D$502,$F7)	7.57
25.0	= PERCENTILE($D$3:$D$502,$F8)	7.71
30.0	= PERCENTILE($D$3:$D$502,$F9)	7.86
35.0	= PERCENTILE($D$3:$D$502,$F10)	8.00
40.0	= PERCENTILE($D$3:$D$502,$F11)	8.14
45.0	= PERCENTILE($D$3:$D$502,$F12)	8.29
50.0	= PERCENTILE($D$3:$D$502,$F13)	8.43
55.0	= PERCENTILE($D$3:$D$502,$F14)	8.43
60.0	= PERCENTILE($D$3:$D$502,$F15)	8.71

Table 4.3 *Concluded*

The percentile	The Excel formula	The percentile value
65.0	= PERCENTILE($D$3:$D$502,$F16)	8.71
70.0	= PERCENTILE($D$3:$D$502,$F17)	8.86
75.0	= PERCENTILE($D$3:$D$502,$F18)	9.00
80.0	= PERCENTILE($D$3:$D$502,$F19)	9.14
85.0	= PERCENTILE($D$3:$D$502,$F20)	9.43
90.0	= PERCENTILE($D$3:$D$502,$F21)	9.59
95.0	= PERCENTILE($D$3:$D$502,$F22)	10.00
Maximum	= MAX($D$3:$D$502)	11.29

And use these results in a RMT/cumul function:

= RMT/cumul (minimum, maximum, percentile values, percentile %s)

for a single signal and then repetitively for the set of ten to be installed (the Excel references are to the function's parameters in an imagined spreadsheet):

The project needs to erect ten signal posts. This is a common task and because it is often done during short closures of the railway, the contractors keep records of how long it takes to do the work. The range of times noted has a broad spread because the works have been carried out at sites that have different access problems and because they have been done in different weather conditions.	= (RMT/cumul (I3,I23,I4:I22,G4:G22) + RMT/cumul (I3,I23,I4:I22,G4:G22) + RMT/cumul (I3,I23,I4:I22,G4:G22) + RMT/cumul (I3,I23,I4:I22,G4:G22) + RMT/cumul (I3,I23,I4:I22,G4:G22) + RMT/cumul (I3,I23,I4:I22,G4:G22) + RMT/cumul (I3,I23,I4:I22,G4:G22) + RMT/cumul (I3,I23,I4:I22,G4:G22) + RMT/cumul (I3,I23,I4:I22,G4:G22) + RMT/cumul (I3,I23,I4:I22,G4:G22))/7

EXAMPLE RISK 17: SPREAD DURATION BASED ON ERROR MEASUREMENT

A project is one-fifth complete, with about 100 tasks finished. It is also running behind schedule, with many tasks taking longer than expected. There is a concern that there has been a degree of optimism in the planning.

- This risk is one that arises in projects that have started. Because the example model is one for a project that has not, an example of this type of primitive is provided here as a stand-alone worksheet – Error Function Model – in the example model workbook. Its outputs are not linked back to the time risk emulation.

I have occasionally modelled time spreads as the biases of planners when they chose the durations of tasks in their plans. It can also be adapted to reveal biases in estimating costs.

Like the previous example of time spreads derived from historical data, some preliminary calculations need to be done. First an overrun ratio is calculated for each completed task in the plan:

overrun ratio = (actual duration – planned duration)/planned duration

An example is shown in Table 4.4.

Table 4.4   Example of a spread duration based on error measurement

On-site contractor's task no.	Planned duration in lapsed weeks	Actual duration in lapsed weeks	Overrun ratio
PH004-1	9.6	9.7	1.0138
PH004-2	1.3	1.2	0.9282
PH004-3	7.0	7.4	1.0568
PH004-4	8.0	8.2	1.0181
PH004-5	9.6	10.5	1.0916
PH004-6	16.0	19.2	1.1988
PH004-7	12.8	12.7	0.9955
PH004-8	12.9	11.8	0.9184
PH004-9	11.8	13.8	1.1697
PH004-10	18.9	19.2	1.0203
PH004-11	5.6	7.0	1.2380
PH004-12	13.9	13.0	0.9390
PH004-13	10.7	10.6	0.9898
PH004-14	3.2	3.3	1.0302
PH004-15	10.7	11.6	1.0855
PH004-16	1.9	2.3	1.1799
PH004-17	4.1	4.1	1.0187
PH004-18	12.5	12.1	0.9724
PH004-19	14.3	16.3	1.1374
PH004-20	14.9	16.1	1.0783
PH004-21	1.8	1.9	1.0285
PH004-22	3.3	3.2	0.9808
PH004-23	6.3	6.1	0.9638
PH004-24	3.0	3.5	1.1553
PH004-25	2.5	2.8	1.1367
PH004-26	11.1	11.7	1.0508
PH004-27	2.4	2.5	1.0564
PH004-28	4.2	4.4	1.0632
PH004-29	17.7	19.5	1.1019
PH004-30	8.9	8.8	0.9855

I then use the RMT distribution fitting tool to find the function that most closely matches the spread of overrun ratios (Figure 4.10).

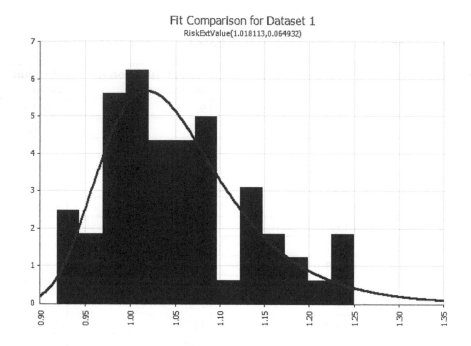

Figure 4.10 A range of overrun ratios

The histogram is the data and the curve is the fitted distribution. In this case it is

= RMT/extvalue (1.018113, 0.064932).

This is the overrun ratio curve and it indicates the spread of errors in the forecasts of task durations to date. Ratios that are less than 1.0 indicate some tasks were finished in less time than planned, though usually there are many more that are greater than 1.0, indicating the tasks that took longer. Overall, the analysis shows there has been a persistent bias towards underestimating how long tasks would take.

If I make an assumption that this optimism was present throughout the process of devising the plan, then the overrun ratio curve probably applies equally to those tasks that have yet to be started (or finished). I would therefore model the durations of these tasks as:

Modelled duration of a task = planned duration * (1 + RMT/extvalue (1.018113, 0.064932)).

An example is given in Table 4.5.

Table 4.5 Using the error function

Task no.	Planned duration in weeks	Error function model	Modelled duration in weeks
PH004-101	2.6	= 2.6 * (1 +RMT/Extvalue (1.018113, 0.064932))	3.8
PH004-102	19.5	= 19.5 * (1 + RMT/Extvalue (1.018113, 0.064932))	30.3
PH004-103	4.8	= 4.8 * (1 + RMT/Extvalue (1.018113, 0.064932))	5.8
PH004-104	10.7	= 10.7 * (1 + RMT/Extvalue (1.018113, 0.064932))	17.3

The modelled durations in the right-hand column will change with each iteration of a simulation, and will affect the overall emulation as before.

If you look in columns G and U of the time sheet of example model, you will see most of the tasks without the specific risks described in this section do have time spreads e.g. U6 is RMT/Uniform (20, 28). These reflect the overruns due to planning optimism for the actual project from which the example model was derived. If, say, I had developed an emulation of a project for which no error measurement was possible yet because it had not started I could calculate the overrun ratio curve and update the model by replacing the fixed durations of the remaining tasks as shown in Table 4.5: Using the Error Function.

Fuzzy Overlap Primitive

EXAMPLE RISK 18: VARIABLE LAG

Piling is scheduled to start four weeks after the surveying and site clearing have started, however the rigs are being used on another project and their arrival on this project could be anytime between four and eight weeks.

- See cells A15 – U15 in the time risk modelling sheet of the example model (Table 4.6).

Table 4.6   Example of a fuzzy overlap

Row/column	K	L	M	P	U
	Task	Start	Predecessor task numbers	Risk	Duration
15	Mobilise the piling contractor and construct piles	= J13 + RMT/ Uniform (4,8)	11SS+4	Piling is scheduled to start 4 weeks after site clearance has started, however the rigs are being used on another project and their arrival on this project could be anytime between 4 and 8 weeks. The piling will then take 6.12 weeks to do depending on productivity rates.	= RMT/Uniform (6,12)

Here, the task of mobilising a piling contractor to construct piles is planned to commence four weeks after the site clearance task has started, not when it has finished, implying both will be underway on the site. The risk is that the contractor may be delayed by up to four more weeks, changing the lag of four into a variable lag of four to eight weeks (Figure 4.11).

It is possible to see the same scenario as the clearance task needing to commence four weeks in advance of the piling contractor – a fixed lead. If the clearance task had a risk of delay of up to a further four weeks, then it might delay the piling by four to eight weeks again. This would be a variable lead. Note that in this case the risk is with the clearing and not the piling, even though the overall effect would be the same. I group both, variable lead and lag, under the same name – fuzzy overlaps.

The lag shown in the example does not extend the duration of the task but simply delays its start. The risk function for the delay, RMT Uniform (4,8) therefore operates

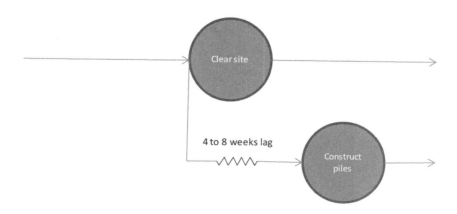

Figure 4.11 A variable lag

on the start of piling task: see cell J15. The start week is read in from the finish week of the preceding task, between four and eight weeks are added to it and then the task itself commences.

Also note that in P15, a spread duration of 6–12 weeks replaces a spot duration of 6 weeks to model the risk that the piling could take longer to do than planned. Thus we now have two risks affecting one activity. More could be added if appropriate.

Here is an important distinction between cost risk modelling and time risk modelling. Cost risk modelling is almost always a monolithic list of risks whereas time risk modelling often is not. You cannot easily read the list of time risks in the example risk model because they are scattered, and sometimes as here, compounded.

This style does cause problems for reviewers and analysts alike when preparing tables for reports, so what I have sometimes done is to replace all the risk content in a time risk model with references to a risk data table in another worksheet that contains a list of the time risks, their analyses and modelling.

The fuzzy overlap risk is very common and can occur in many situations. For example:

- Waiting for plaster to dry before painting.
- Waiting for concrete to set before attaching fixings.
- Waiting for plant to cool before opening it up.
- Waiting for pollutant levels to decay before going in.
- Waiting for resources to turn up.

The Emerging and Disappearing Task, or Logic, Primitive

EXAMPLE RISK 19: SIMPLE EMERGING TASK

There is a possibility that some landscaping work will have to be done to make the site acceptably attractive.

- See cells A48 – U48 in the time risk modelling sheet of the example model (Table 4.7).

Table 4.7   Example of emerging logic: 1

Row/column	J	K	P	Q	R	S	T	U
	Task	**Start**	**Risk**				**Duration**	**Finish**
48	Landscape the site	= Q26	There is a possibility that some landscaping work will have to be done to make the site acceptably attractive.	Emerging task. Very likely to be a 75% chance of another 1.4 weeks.	= RMT/ Discrete ({1,0}, {75,25})	= RMT/ Uniform (1,4)	= N27 * O27	= J27 + P27

In the example model the final activity is to clear the site but there is a risk. It will have to be landscaped to be acceptable to those who live next to it. This is an emerging task.

It is built into the emulation in the usual manner with a start link, a duration and an end calculation, but whereas on the spot side of the model the duration is set to 0 weeks (see E48), so the task has no effect on the overall duration of the plan. On the risk side (see R48) it acquires 75 per cent chance of needing 1–4 weeks and 25 per cent chance of none. When the model is run, on some iterations the task pops up with duration somewhere between one and four weeks, i.e. it emerges, while on other occasions it has no effect at all.

It is easy to see that several tasks in a sequence could be modelled as emergent if they all linked together in the emulation but were subject to the same logical test as to whether their durations should be finite (the sequence has emerged) or zero (the sequence has not emerged).

It is also easy to see the opposite can be modelled. A task or a sequence of tasks confirmed as being in the plan can be made to disappear from the emulated logic by setting the durations to zero. This is only necessary if an opportunity is discovered that they need not be done.

EXAMPLE RISK 20: COMPOUND EMERGING TASK

Emerging tasks are very common in time risk modelling. Here is a sequence where the project will complete one task or another but not both, in contrast to one task or nothing. It is easy to see that these mutually exclusive tasks could be mutually exclusive sequences of tasks.

- See cells A18 – Q20 in the time risk modelling sheet of the example model (Table 4.8).

Table 4.8   Example of emerging logic: 2

	Task	**Risk**	**P**	**Q**	**R**	**S**	**U** Duration
18	Do general civils	Public consultation might result in need to put up fencing …	Emerging task: very likely so 75% chance of another 4.6 weeks work. Note the costs are in the estimate.	= Risk Discrete ({0,1}, {25,75})	= Risk Uniform (4,6)		= E18 + (N18 *IF(O18 = 1, O18,O19)) + O20
19		.... or public consultation might result in need to construct noise bunds.	emerging task: very likely but as an alternative to fencing on 50/50 basis. Another 6.10 weeks. No extra costs incurred.	= Risk Discrete ({1,0}, {50,50})	= Risk Uniform (6,10)		
20		Risk 45 from the cost model	Emergent cost chain: 20% chance of a slip during the duration of the project, incurring 250–500k of works and 4–6 week extension.	= 'cost risk model'! F41	= CHOOSE (N20 + 1,0, RMT/Uniform (4,6), RMT/Uniform (4,6) + RMT/Uniform (4,6), RMT/Uniform (4,6) + RMT/Uniform (4,6) + RMT/Uniform (4,6), RMT/Uniform (4,6) + RMT/Uniform (4,6) + RMT/Uniform (4,6) + RMT/Uniform (4,6))	

Here the task of carrying out the civil works is at risk of one of two emerging tasks that have to be done as a consequence of public consultation. Either a fence or earth bunds may have to be constructed around the site to reduce the construction noise, or possibly not because the noise levels are acceptable.

There is one further risk and this is partly modelled in the cost part of the Example Model: Landslips (see A66 : I75 of the cost risk modelling sheet from the example model) (Table 4.9).

Table 4.9   The landslip example from the cost model

Row/column	A	B	C
41	7	The embankment to the west has a history of slips: one in the last five years. Another slip could happen during the works.	emerging cost chain: 20% chance of a slip during the duration of the project, incurring £250–£500k of *works and 4–6 week extension*

This risk has both a cost and time risk impact. As you can see in cell R20 of Table 4.8, the time risk model acquires the number of landslips that may occur from Risk 6 in the cost model, and computes the time required to deal with them in a chain of accumulated RMT Uniform functions indexed by an Excel CHOOSE function. Note that as in the cost model, each landslip is treated independently in respect of the amount of time required to recover from its effects.

The accumulated effect of these emerging tasks is computed in cell U18 and added to the original spot duration from E18.

Rework Cycle Primitive

EXAMPLE RISK 21: REWORK CYCLE

The chosen type of non-interruptible back-up generators may fail their full load tests. If they do, they will have to be up-rated by reinstating features that were removed by a Value Engineering exercise. The commissioning task will have to be repeated if this happens. In the opinion of the electrical engineers, the removed features were important to meeting availability, maintainability and reliability (ARM) targets so there is a 50 per cent chance of a recommissioning.

- See cells A23 – V23 in the time risk modelling sheet of the example model (Table 4.10).

Table 4.10 Example of a rework cycle

Row/column	J	P	Q	U	V
	Start	**Risk**	**Modelling**	**Duration**	**Finish**
23	= MAX (Q17, Q18, Q21, Q22)	The chosen type of non-interruptible back-up generators may fail their full load tests. If they do they will have to be up-rated by reinstating features that were removed by in a Value Engineering exercise. The commissioning task will have to be repeated if this happens. In the opinion of the electrical engineers there is a 50% chance of this happening.	Rework cycle: 50% chance that the commissioning will have to be repeated	= E23 + (RMT/Discrete ({0,1},{50,50})) * E23	= J23 + P23

A rework cycle is a very common occurrence where tests have to be carried out, such as in the above example, or an application for authorisation to spend is made or planning permission is sought. The risk here is that the work will have to be redone, or the application resubmitted because the authorities have rejected it.

Here, a rework cycle manifests itself. Time may be needed to repeat the commissioning tests on a set of non-interruptible power supplies. These have been de-specified as a result of a value engineering exercise that (presumably) someone the analyst has interviewed has doubted the wisdom of doing.

The modelling is identical to that for an emerging task – an additional amount of time may need to be added to the spot duration needed (see cell U23, in which two weeks may be added at a 50 per cent probability to extend the spot duration).

EXAMPLE RISK 22: PART REWORK CYCLE

Sometimes a rework cycle can be a part rework cycle. In the following example, for instance, the task is not fully repeated.

Due to a tightening of authorisation levels, contracts may be rejected. This will mean that the limits of delegated authority will be reduced and so one further higher level signatures may be needed. This will not require an entire reapplication, however, and a signed memorandum confirming the approval will suffice.

- See cells A11 – V11 in the time risk modelling sheet of the example model (Table 4.11).

Table 4.11 Example of a part rework cycle

Row/column	J	P	Q	U	V
11	= Q10	An application for authorisation to let contracts may be rejected because a tightening of authorisation levels is anticipated. This will mean that the limits of delegated authority will be reduced and so one further higher level signature may be needed. This will not require an entire reapplication, and a signed memorandum confirming the approval will suffice.	Part-rework cycle: 10% chance of an additional 25% of the planned time frame of the contract negotiation period.	= E11 + (RMT/Discrete ({1,0}, {10,90}) * (E11*25%))	= J11 + P11

Since the modelling is the same for the rework cycle as that of the emerging task, i.e. an additional amount of time may or may not be added to the original duration, the reasonable question of why not call them the same thing, say the Additional Time Primitive, may well pop up. The reason is that the primary purpose of primitives is to facilitate risk identification. Sometimes fishing for risks is more successful when emerging tasks are the bait, and at other times when rework cycles are. In my experience, people from the construction community tend to foresee emerging tasks because they anticipate things going wrong, while those from the development community dealing with consultations, negotiations and consents tend to foresee rework cycles because they have endured many past rejections of their efforts.

There is however one variation of the rework cycle that does give the primitive some distinction: the multiple rework cycle.

EXAMPLE RISK 23: MULTIPLE REWORK CYCLE

The application for funds to purchase new plant is likely to be rejected repeatedly, each rejection taking the form of a request for further information. The investment committee sits every four weeks. It has been decided to plan for three reapplications on an assumption a fourth one will be successful. There is thought to be a 25 per cent chance of success with the first application, 50 per cent with a second, 15 per cent with a third and 10 per cent with a fourth.

- See cells A9 – V9 in the time risk modelling sheet of the example model (Table 4.12).

Table 4.12   Example of a multiple rework cycle

Row/column	J	P	Q	U	V
9	= Q8	The application for funds to purchase new plant is likely to be rejected repeatedly, each rejection taking the form of a request for further information. The investment committee sits every 4 weeks. It has been decided to plan for three reapplications on an assumption a fourth one will be successful. There is thought to be a 25% chance of success with the first application, 50% with a second, 15% with a third and 10% with a fourth.	Multiple rework cycle: 25% chance of one cycle (4 weeks); 50% chance of two cycles; 15% of three and 10% of four.	= RMT/Discrete ({4,8,12,16}, {25,50,15,10})	= J9 + P9

Risks of this type are common in situations where there are objections to the project that must be overcome by repeated requests for permission, and in situations where audits and inspections have to be repeated in order to clear a list of deficiencies. In the example, an application for funding is repeatedly rejected.

Like the previous two examples, the risk is modelled as an additional period of time that may be added to the duration. Except on this occasion there are multiple possible durations reflecting the possible occurrence of repeated rework cycles.

The formula for the risk is:

Duration = ((RMT/Discrete{4 weeks, 8 weeks, 12 weeks, 16 weeks}, {25%,50%,15%, 10%})

I have put the spot duration of four weeks directly into the formula because the RMT Discrete function will not accept cell references as parameters – if the spot duration ever changes I need to change the formula accordingly. I highlight cells that have this limitation to remind me.

Interruption Primitive

EXAMPLE RISK 24: INTERRUPTION

Installation engineers booked for this task may have to be diverted to a major system commissioning task on project X that is also planned to happen at this time.

- See cells A22 – V22 in the time risk modelling sheet of the example model (Table 4.13).

Table 4.13   Example of an interruption

Row/column	J	P	Q	U	V
22	= Q16	Installation engineers booked for this task may have to be diverted to a major systems commissioning task on project X that is also planned to happen at this time.	Interruption: low probability, but project X will take priority on resources. If it does then this commissioning will have to be deferred two weeks.	= E22 + RMT/Discrete ({1,0}, {25,75}) * 2	= J22 + P22

Interruptions are events that stop scheduled work, as distinct from disruptions which are events that reduce but don't altogether stop work happening. In the example, work may be suspended because another project has a higher priority claim on skilled staff, and as for the preceding examples, the consequence is to add a period of time to the spot duration.

EXAMPLE RISK 25: THE MULTIPLE TASK INTERRUPTION

There is a type of interruption that is modelled differently, however. In the preceding example the interruption affected a single task within the emulation, but some discrete interruptions may affect several tasks if they are scheduled to be taking place at the same time. For example:

All construction activities may be suspended if there is a major security alert.

Or

Bad weather may prevent staff getting to the site.

In these cases the risk is not a characteristic of the task but of the time period in which the task may be underway. One can conceptualise a number of concurrent tasks, which because of the logic in their underlying plan, end up being scheduled for wintertime (more prone to bad weather). The risk is to the working calendar and not the tasks themselves.

In Figure 4.12, a group of tasks is displaced three days by the unwanted outcome of the risk of a storm.

This is very tricky to compute in an emulation because the start and finish dates of the discrete tasks are determined by their own risks. These risks that affect their logical predecessors in the plan change from model iteration to iteration. Therefore, it is generally not possible to say which tasks will comprise the group affected by the many-task interruption and so possibly add an additional period of time to each of them to model the risk.

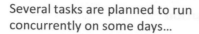

Several tasks are planned to run
concurrently on some days...

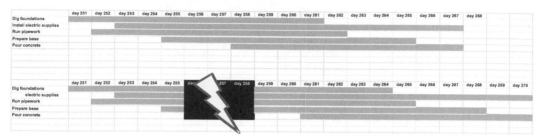

...but a three day storm interrupts them

Figure 4.12 A many-task interruption

The answer is to model the risk within the calendar of working days pertaining to each task. Such a calendar will typically be a five-, six- or a seven-day working week so often there will be just one or two calendars that are common to all the tasks in a plan. There can of course be subtleties such as shift work, different holiday patterns and weekend only working.

The risk has to knock out working days in a calendar, by which I mean change them from being a working to a non-working day. The emulation has therefore to be expanded to include calendars and by doing so it will become much more difficult to understand. I, therefore, limited the interrupted calendar to one task subject and one risk to show you how the risk can be modelled. If I were to extend this modelling to cover every task and every calendar (even if it was shared with other tasks) then the model would become too dense and intractable for the purposes of explanation.

In a sheet called 'five-day calendar' in the example model workbook, a column of calendar days that amply cover the duration of the project is set up in column A (see Table 4.14 for values and Table 4.15 for formulae). The pattern of 1s and 0s in column C denotes the working days as 1, and non-working days as 0; this is based on a five-day working week calendar. The pattern of 1s and 0s in column D similarly distinguishes January and February from the other months. A logical test in column E marks select working days in January and February but this is overwritten by an RMT function that will force a working day to be a non-working day when the temperature falls below zero. The same table with formulae displayed follows. Conditional shading is used to highlight the patterns of 1s and 0s.

An accumulation of the working days available after the modelling of the ambient temperature risk is placed in column F, and in column G, the date from column A is repeated (Tables 4.14 and 4.15).

Values:

Table 4.14 Example of calendar risk: values

Row/column	A	B	C	D	E	F	G
1	Date	Day	Test for working days in the week	Test for January or February	Risk: a working day in either January or February on which the ambient temperature falls below 0°C becomes a non-working day.	Total working days after risk is modelled	Date (repeat)
2	01-Jan-12	Sun	0	1	0	0	01-Jan-12
3	02-Jan-12	Mon	1	1	1	1	02-Jan-12
4	03-Jan-12	Tue	1	1	1	2	03-Jan-12
5	04-Jan-12	Wed	1	1	1	3	04-Jan-12
6	05-Jan-12	Thu	1	1	1	4	05-Jan-12
7	06-Jan-12	Fri	1	1	1	5	06-Jan-12
8	07-Jan-12	Sat	0	1	0	5	07-Jan-12
9	08-Jan-12	Sun	0	1	0	5	08-Jan-12
10	09-Jan-12	Mon	1	1	1	6	09-Jan-12
11	10-Jan-12	Tue	1	1	1	7	10-Jan-12
12	11-Jan-12	Wed	1	1	1	8	11-Jan-12
13	12-Jan-12	Thu	1	1	1	9	12-Jan-12

and so on until 31 December 2014.

Formulae:

Table 4.15 Example of calendar risk: formulae

Row/column	A Date	B Day	C Test for working days in the week	D Test for January or February	E Risk: a working day in either January or February on which the ambient temperature falls below 0°C becomes a non-working day.	F Total working days after risk is modelled.	G Date (repeat)
1	Date	Day	Test for working days in the week	Test for January or February	Risk: a working day in either January or February on which the ambient temperature falls below 0°C becomes a non-working day.	Total working days after risk is modelled.	Date (repeat)
2	= DATE (2012,1,1)	Sun	= IF (AND(WEEKDAY(A2)<>1,W EEKDAY(A2)<>7),1,0)	= IF (OR(MONTH(A2)= 1,MONTH(A2)=2),1,0)	= IF (AND(C2=1,D2=1),RMT/ Binomial(1,90%),C2)	= E2	= A2
3	= A2 + 1	Mon	= IF (AND(WEEKDAY(A3)<>1,W EEKDAY(A3)<>7),1,0)	= IF (OR(MONTH(A3)= 1,MONTH(A3)=2),1,0)	= IF (AND(C3 = 1,D3 = 1),RMT/ Binomial(1,90%),C3)	= F2 + E3	= A3
4	= A3 + 1	Tue	= IF (AND(WEEKDAY(A4)<>1,W EEKDAY(A4)<>7),1,0)	= IF (OR(MONTH(A4)= 1,MONTH(A4)=2),1,0)	= IF (AND(C4 = 1,D4 = 1),RMT/ Binomial(1,90%),C4)	= F3 + E4	= A4
5	= A4 + 1	Wed	= IF(AND(WEEKDAY(A5)<>1,W EEKDAY(A5)<>7),1,0)	= IF (OR(MONTH(A5) = 1,MONTH(A5) = 2),1,0)	= IF (AND(C5 = 1,D5 = 1),RMT/ Binomial(1,90%),C5)	= F4 + E5	= A5
6	= A5 + 1	Thu	= IF (AND(WEEKDAY(A6)<>1,W EEKDAY(A6)<>7),1,0)	= IF (OR(MONTH(A6) = 1,MONTH(A6) = 2),1,0)	= IF (AND(C6 = 1,D6 = 1),RMT/ Binomial(1,90%),C6)	= F5 + E6	= A6
7	= A6 + 1	Fri	= IF (AND(WEEKDAY(A7)<>1,W EEKDAY(A7)<>7),1,0)	= IF (OR(MONTH(A7) = 1,MONTH(A7) = 2),1,0)	= IF (AND(C7 = 1,D7 = 1),RMT/ Binomial(1,90%),C7)	= F6 + E7	= A7
8	= A7 + 1	Sat	= IF (AND(WEEKDAY(A8)<>1,W EEKDAY(A8)<>7),1,0)	= IF (OR(MONTH(A8) = 1,MONTH(A8) = 2),1,0)	= IF (AND(C8 = 1,D8 = 1),RMT/ Binomial(1,90%),C8)	= F7 + E8	= A8
9	= A8 + 1	Sun	= IF (AND(WEEKDAY(A9)<>1,W EEKDAY(A9)<>7),1,0)	= IF (OR(MONTH(A9) = 1,MONTH(A9) = 2),1,0)	= IF (AND(C9 = 1,D9 = 1),RMT/ Binomial(1,90%),C9)	= F8 + E9	= A9
13	= A12 + 1	Thu	= IF (AND(WEEKDAY(A13)<>1, WEEKDAY(A13)<>7),1,0)	= IF (OR(MONTH(A13) = 1,MONTH(A13) = 2),1,0)	= IF (AND(C13 = 1,D13 = 1),RMT/ Binomial(1,90%),C13)	= F12 + E13	= A13

The risk looks like this:

- See cells A46 – V46 in the time risk modelling sheet of the example model (Table 4.16).

Table 4.16 Example of lost working days

Row/column	J	P	Q	R	S	T	U	V	W	X
	Start	Synopsis	Modelling note	#1	#2	Duration	Finish	Start date	Finish date	
46	= R46	The evaluation period requires 4 weeks of ambient temperatures >0°C. If the task happens to fall in January or February there is a 10% chance that any working day will experience a temperature lower than this, in which case an additional day will need to be added to the overall duration of the task.	Modelled as a 10% chance of any working day being lost during January and February in 5-day calendar.	= VLOOKUP (S46, '5 day calendar' !$A$2:$G$1097, 6,FALSE)	= N46 + (E46 * 7)	= VLOOKUP (O46, '5 day calendar' !$F$2:$G$1097, 2,FALSE)	= (T46 - S46) /7	= J46 + Q46	= INT ($G$3 + J46*7)	= P46

The start date of the task in cell V46 is made equal to the start date of the project plus the number of days since the project started, J46.

The question is when does the task finish, given that its duration may have to be extended if the task has to be done in January or February? This is done by looking up the start date in the five-day calendar, N25, and reading from the calendar the total number of working days in the project up to that date. This total is made after the potential loss of working days due to temperatures being below zero has been modelled. This will always be equal to, or fewer than, the number of working days that notionally existed before the risk was modelled.

In S46, the spot duration is converted from weeks to days by multiplying by seven, and is added to the value of the preceding Excel Lookup. This value is used in P46 to search down the calendar for the finish date using a second Excel Lookup. The number of steps in the search down will be exactly equal to the spot duration in days if none have been lost, or equal to the spot, plus the number lost, when some have. This is because lost days are marked as 0s in column E of the calendar and so do not alter the total working days in column F of the calendar.

The finish date appears in P46 and for completeness is copied to T46. The possibly extended duration in weeks is calculated in Q46 for completeness.

This is a powerful primitive because it moves the risk from the work to the context for the work – the calendar – and further, by modelling the risk in the calendar and not the task, the risk can be used to affect more than a single task. It never rains but it pours. Make the analysis fit the problem.

Disruption Primitive

EXAMPLE RISK 26: DISRUPTION

The clients may wish to move onto the site two weeks after the evaluation period, in case the last half of the time required to clear the site will be disrupted and take up to 100 per cent longer (Table 4.17).

Table 4.17 Example of disruption

Row/ column	J	P	Q	U	V
	Start	**Synopsis**	**Modelling note**	**Duration**	**Finish**
47	= R25	The clients may wish to move onto the site two weeks after the evaluation period, in case the last half of the time required to clear the site will be disrupted and take up to 50% longer.	Disruption after 50% completion, with up to 100% extension of the remaining balance of time.	= (E47*50%) + (E47*(1-50%) * (1+RMT/Uniform (0%,100%)))	= J47+Q47

- See cells A27 – V27 in the time risk modelling sheet of the example model.

A disruption is an event that alters a rate of production or progress at some point in time but does not stop it altogether. In the above example, clients wish to start to move into a site before it has been cleared. The risk is that their doing so will disrupt progress on this task. Here, the spot duration of the clearance task is four weeks and the disruption event will happen, if it does at all, after two weeks. The balance of the clearance work will then take twice as long to do.

In the modelling, the risked duration is calculated as follows:

- (E47 * 50%) calculates the length of time for which full production was possible, which is added to
- E47 * (1–50%), the balance of time that will be subjected to the disruption after this has been extended by up to 100%, 1 + RMT/Uniform (0%, 100%).

I have not written them up as a separate examples but it is easy to see.

The point of disruption could itself be a variable instead of a fixed parameter i.e. instead of the halfway point, 50 per cent, above, it could be

= RMT/Uniform (start week, finish week) – start week/(finish week – start week)

= somewhere between 0% and 100% of the way through the planned duration.

This would allow the risk to be modelled as a disruption caused by an external event occurring at some random time during the task.

By replacing the Uniform function with, say, a Triangular one, the timing of the occurrence of an external event could be biased to be more likely to happen at some particular stage of the task: early on, or midway, or towards the close. A good example of this would be the tendency for disruption by protestors to be intense at the start of controversial projects, but ameliorate as time passes. Also..

The multiple task interruption modelling that knocked working days out of the working day calendar could be adapted to model multiple task disruption. This can be achieved by counting not 1s and 0s to represent working and non-working days respectively, but 100 per cent and X per cent, where X per cent is somewhere between 0 per cent and 100 per cent to denote full and disrupted days respectively. Instead of the accumulation search algorithm looking for say, 10 working days, it would have to seek 10 * 100% = 10000%s worth of working time.

Productivity Primitive

EXAMPLE RISK 27: VARIABLE PRODUCTIVITY

Reinstatement of the overhead wiring is planned to take place over 10 days – 20 sections at a rate of 2 sections per day. The historical record shows that the initial period of work is more susceptible to plant breakdowns and logistical mishandlings, but that productivity does improve to a sustained average of four sections per day (Table 4.19).

Table 4.18 Example of productivity growth

Row/column	Task	Start	Synopsis	Modelling note	#1	#2	#3	Duration	Finish
	K	L	N	O	P	Q	R	S	T
24	re-instate over-head wiring	= T23	Reinstatement of the overhead wiring is planned take place over 10 days: 20 sections at a rate of 2 sections per day. The historical record shows that early says that early says are susceptible to plant breakdowns and logistical mishandlings but that productivity does improve to a sustained average of 4 sections per day.	Productivity growth: each day is modelled with production rate of 0 to 6 sections but the probability of each outcome biased towards 0 for the first working week, thereafter altering steadily towards 6 sections.				= VLOOKUP (1,R27:S46,2,FALSE) /7	= L25 + S25
25				Day no. label	Production	Acc. Production	Quota achieved?	Day no.	
26				Day 1	= RMT/ Discrete ({0,1,2,3,4,5,6}, {40,30,20,10,0,0,0})	= P27	= IF (Q27<20, 0,1)	1	
27				Day 2	= RMT/ Discrete ({0,1,2,3,4,5,6}, {40,30,20,10,0,0,0})	= Q27 + P28	= IF(Q28<20,0,1)	2	
28				Day 3	= RMT/ Discrete ({0,1,2,3,4,5,6}, {30,40,20,10,0,0,0})	= Q28 + P29	= IF(Q29<20,0,1)	3	
29				Day 4	= RMT/ Discrete ({0,1,2,3,4,5,6}, {20,40,30,10,0,0,0})	= Q29 + P30	= IF(Q30<20,0,1)	4	
30				Day 5	= RMT/ Discrete ({0,1,2,3,4,5,6}, {20,30,40,10,0,0,0})	= Q30 + P31	= IF(Q31<20,0,1)	5	

Table 4.18 Concluded

31	Day 6	= RMT/ Discrete ({0,1,2,3,4,5,6}, {10,30,40,20,0,0,0})	= Q31 + P32	= IF(Q32<20,0,1)	6
32	Day 7	= RMT/ Discrete ({0,1,2,3,4,5,6}, {10,20,40,20,10,0,0})	= Q32 + P33	= IF(Q33<20,0,1)	7
33	Day 8	= RMT/ Discrete ({0,1,2,3,4,5,6}, {10,10,30,30,10,10,0})	= Q33 + P34	= IF(Q34<20,0,1)	8
34	Day 9	= RMT/ Discrete ({0,1,2,3,4,5,6}, {10,10,30,20,10,10,10})	= Q34 + P35	= IF(Q35<20,0,1)	9
35	Day 10	= RMT/ Discrete ({0,1,2,3,4,5,6}, {10,10,30,20,10,10,10})	= Q35 + P36	= IF(Q36<20,0,1)	10
36	Day 11	= RMT/ Discrete ({0,1,2,3,4,5,6}, {10,10,30,20,10,10,10})	= Q36 + P37	= IF(Q37<20,0,1)	11
37	Day 12	= RMT/ Discrete ({0,1,2,3,4,5,6}, {10,10,30,20,10,10,10})	= Q37 + P38	= IF(Q38<20,0,1)	12
38	Day 13	= RMT/ Discrete ({0,1,2,3,4,5,6}, {10,10,30,20,10,10,10})	= Q38 + P39	= IF(Q39<20,0,1)	13
39	Day 14	= RMT/ Discrete ({0,1,2,3,4,5,6}, {10,10,30,20,10,10,10})	= Q39 + P40	= IF(Q40<20,0,1)	14
40	Day 15	= RMT/ Discrete ({0,1,2,3,4,5,6}, {10,10,30,20,10,10,10})	= Q40 + P41	= IF(Q41<20,0,1)	15
41	Day 16	= RMT/ Discrete ({0,1,2,3,4,5,6}, {10,10,30,20,10,10,10})	= Q41 + P42	= IF(Q42<20,0,1)	16
42	Day 17	= RMT/ Discrete ({0,1,2,3,4,5,6}, {10,10,30,20,10,10,10})	= Q42 + P43	= IF(Q43<20,0,1)	17
43	Day 18	= RMT/ Discrete ({0,1,2,3,4,5,6}, {10,10,30,20,10,10,10})	= Q43 + P44	= IF(Q44<20, 0,1)	18
44	Day 19	= RMT/ Discrete ({0,1,2,3,4,5,6}, {10,10,30,20,10,10,10})	= Q44 + P45	= IF (Q45<20, 0,1)	19
45	Day 20	= RMT/ Discrete ({0,1,2,3,4,5,6}, {10,10,30,20,10,10,10})	= Q45 + P46	= IF (Q46<20, 0,1)	20

- See cells A24 – T45 in the in the time risk modelling sheet of the example model.

Productivity growth occurs for programme-type work, which is to say work repeated day after day (a project has an end whereas a programme does not. The work simply continues. A group of projects is most often referred to as a programme, especially where as one project is finished, another is added). In the example, 20 sections of overhead power cables have to be reinstated. The spot duration has been determined by the planners as two weeks on a conservative assumption of two sections per working day. In practice though, the production rates of repeated tasks tend to speed up once the work site and the working methods become familiar to those doing the work.

In a lengthy calculation that extends over a number of rows, an expected growth in productivity has been modelled as follows. A range of daily productions of between 0 and 6 sections has presumably been discovered in the historical records of such work (which, being done by contractors who are ever on the lookout for claims opportunities, will exist somewhere) and modelled using a RMT discrete function (Table 4.20). However, if you look at the detail of the second parameter of the function, the relative probabilities, you will see that they are biased towards 0 sections being reinstated in the early days of the project, but that this bias shifts towards higher production rates as the days go by.

Table 4.19   Accumulated production

Day no. label	Production	Acc. Production
Day 1	= RMT/ Discrete ({0,1,2,3,4,5,6}, {40,30,20,10,0,0,0})	= P27
Day 2	= RMT/ Discrete ({0,1,2,3,4,5,6}, {40,30,20,10,0,0,0})	= Q27 + P28
Day 3	= RMT/ Discrete ({0,1,2,3,4,5,6}, {30,40,20,10,0,0,0})	= Q28 + P29
Day 4	= RMT/ Discrete ({0,1,2,3,4,5,6}, {20,40,30,10,0,0,0})	= Q29 + P30
Day 5	= RMT/ Discrete ({0,1,2,3,4,5,6}, {20,30,40,10,0,0,0})	= Q30 + P31
Day 6	= RMT/ Discrete ({0,1,2,3,4,5,6}, {10,30,40,20,0,0,0})	= Q31 + P32
Day 7	= RMT/ Discrete ({0,1,2,3,4,5,6}, {10,20,40,20,10,0,0})	= Q32 + P33
Day 8	= RMT/ Discrete ({0,1,2,3,4,5,6}, {10,10,30,30,10,10,0})	= Q33 + P34
Day 9	= RMT/ Discrete ({0,1,2,3,4,5,6}, {10,10,30,20,10,10,10})	= Q34 + P35
Day 10	= RMT/ Discrete ({0,1,2,3,4,5,6}, {10,10,30,20,10,10,10})	= Q35 + P36
Day 11	= RMT/ Discrete ({0,1,2,3,4,5,6}, {10,10,30,20,10,10,10})	= Q36 + P37
Day 12	= RMT/ Discrete ({0,1,2,3,4,5,6}, {10,10,30,20,10,10,10})	= Q37 + P38
Day 13	= RMT/ Discrete ({0,1,2,3,4,5,6}, {10,10,30,20,10,10,10})	= Q38 + P39
Day 14	= RMT/ Discrete ({0,1,2,3,4,5,6}, {10,10,30,20,10,10,10})	= Q39 + P40
Day 15	= RMT/ Discrete ({0,1,2,3,4,5,6}, {10,10,30,20,10,10,10})	= Q40 + P41
Day 16	= RMT/ Discrete ({0,1,2,3,4,5,6}, {10,10,30,20,10,10,10})	= Q41 + P42
Day 17	= RMT/ Discrete ({0,1,2,3,4,5,6}, {10,10,30,20,10,10,10})	= Q42 + P43
Day 18	= RMT/ Discrete ({0,1,2,3,4,5,6}, {10,10,30,20,10,10,10})	= Q43 + P44
Day 19	= RMT/ Discrete ({0,1,2,3,4,5,6}, {10,10,30,20,10,10,10})	= Q44 + P45
Day 20	= RMT/ Discrete ({0,1,2,3,4,5,6}, {10,10,30,20,10,10,10})	= Q45 + P46

A daily accumulated total of the sections reinstated so far is computed and the number of days of production modelled is sufficient for this total to exceed the 20 that do have to be reinstated.

A logical test in column T picks out the day at which the quota is achieved. A simple Excel lookup in U24 scans the test results for the transition point and reads off the number of lapsed working days.

There is a subtlety in U24 that is worth noting. Although the planners assumed a five-day working, the analyst is clearly aware this sort of work is done on a seven-day week calendar and has calculated the number of lapsed weeks using a /7 and not a /5 conversion. Work that is actually done on a seven-day calendar is planned on a five-day calendar and is a regular opportunity to reduce overall timescales. If it is not identified in the risk, and hence brought to the attention of the project management, then Parkinson's Law will tend to apply once the work commences – it will expand to fit the time available by an extra two days in every five.

The Impacts of a Single Risk and of Several Risks in Time Risk Modelling

So far I may have given the impression that one risk has a possible effect on the duration of one task, but it is often the case a task could be affected by the outcomes of several risks.

This can be in one of two broad ways – either the unwanted outcomes of the risks accumulate one after the other so that the extensions of time simply keep pushing the finish date out and out, or they act concurrently so that they cancel each other out to an extent, leaving only one risk (in terms of the extension of time it adds) to determine the duration of the task.

Concurrent: Mutually Exclusive

In the following example (Figure 4.13), two risks that could potentially affect the task duration at the same time – concurrently – do so but in a mutually exclusive way so that the outcome of only one of them prevails.

- See cells A18 – T20 in the in the time risk modelling sheet of the example model.

In Cells N18 and N19 two risks to completion of the civil engineering works are described:

1. Consultation with the public may result in the scope of the works being changed to include either the erection of fencing around the work site *or*
2. The building of earth banks around the work site. This is to both hide it, and provide a measure of construction noise mitigation.

A test is made in P18 using a RMT/Discrete function to determine if an unwanted outcome has materialised; a further test in P19 determines which it is. The possible duration extensions are modelled in Q18 and Q19 and the logic in S18 adds the prevailing risk to the spot duration from E18. Thus the task has two risks but only one impact (Table 4.21).

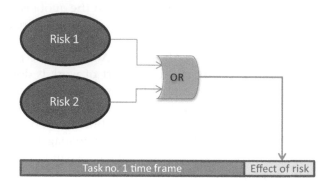

Figure 4.13 Two mutually exclusive risks affecting the same task

Table 4.20   Several risks, one task

Row/ column	K	L	N	O	P	Q	S	T
	Task	Start	Synopsis	Modelling note	#1	#2	Duration	Finish
18	Do general civils	= T16	Public consultation might result in the need to put up fencing....	Emerging task: very likely so 75% chance of another 4..6 weeks work. Note the costs are in the estimate.	= Risk Discrete ({0,1}, {25,75})	= Risk Uniform (4,6)	= E18+ (P18* IF(Q18 = 1,Q18,Q19)) +Q20	= L19 + S19
19	-		... or public consultation might result in need to construct noise bunds.	Emerging task: very likely but as an alternative to fencing on 50/50 basis. Another 6..10 weeks. No extra costs incurred.	= Risk Discrete ({1,0}, {50,50})	= Risk Uniform (6,10)		
20	-		Risk 7 from the cost model	Emerging cost chain: 20% chance of a slip during the duration of the project, incurring 250k...500k of works and 4...6 week extension.	= 'cost risk model'!F41	= CHOOSE (P21 + 1,0, Risk Uniform (4,6), Risk Uniform (4,6) + Risk Uniform (4,6), Risk Uniform (4,6) + Risk Uniform (4,6) + Risk Uniform (4,6), Risk Uniform (4,6) + Risk Uniform (4,6) + Risk Uniform (4,6) + Risk Uniform (4,6))		

Accumulated Unweighted Sum

One further risk appears in N20. This is the landslip risk from the cost model in which the number of slips that could occur is modelled using a RMT/Poisson function. There, the costs were modelled by giving each slip its own independent cost range of between £250k and £500k. Here, the time required is modelled as a number of discrete four to six-week chunks in Q20, the number of which is read from the cost model in P20 and picked out using an Excel choose function (Figure 4.14).

This is an accumulated unweighted sum of the discrete outcomes of each risk that happens. It usually occurs when these are separated in time. In the example, since the time frame of the fencing/earth bank risks is intuitively before that of the landslips risk (the resulting work is likely to be carried out before the main works get underway), I have accumulated the outcomes of both in cell S18: there is potentially a 'double whammy'.

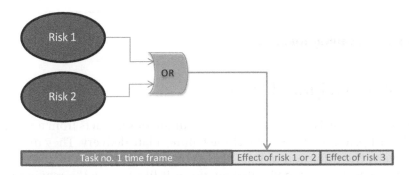

Figure 4.14 Accumulated unweighted sum

Concurrent: Prevailing Maximum

If the planners contradicted my intuition that the fencing/earth banks would be started at the same time as the earthworks (perhaps to keep the overall timescales to a minimum), then I could make the analysis fit the problem by using an Excel maximum function to decide which of them determined the duration of the task, viz (Figure 4.15):

Duration of task = spot duration + maximum

(prevailing duration of the fencing or earth banks risks),

duration of the landslips risk).

The unwanted outcomes of the other risks are thus deemed to be contained within the time frame of the one generating the maximum impact. To avoid confusion, or perhaps accidently to create it, this is not the *biggest* risk but the risk that on the current iteration of the model throws up the largest value. If a small risk throws up a high-end value for the additional chunk of time needed, then the small risk will prevail and set the extension of time for the task both risks affect.

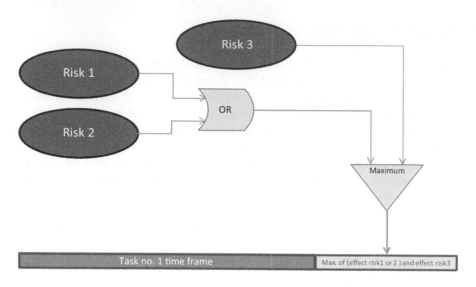

Figure 4.15 A prevailing maximum

Accumulated: Weighted Sum

In some situations a task is exposed to many simultaneous impacts from a set of risks. For example a landslip, a snowstorm, a machine failure, a late delivery. They may all occur at once. Intuitively, the extension of time caused by the risk will not be the accumulation of each discrete event's unwanted outcome, nor is it likely to be the maximum of them, there is clearly a lot more emerging work to be done. Instinctively, the amount of time needed to overcome the unwanted outcomes is somewhere between the maximum risk event and the unweighted sum of all of them.

I use a weighted sum (Figure 4.16) to decide the additional time needed and though it is unproven, I find the 'square root of the sum of the squares' (of the discrete unwanted outcomes) has the right feel to project to people the amount of time needed to deal with multiple conflicting, unwanted outcomes.

Table 4.21   Square root of the sum of the squares

	Sampled unwanted outcome in days	Square root of the sum of the squares	Weighted duration in days
Risk 1	5	$= \sqrt{(5^2=)}$	5.0
Risk 2	6	$= \sqrt{(5^2 + 6^2)}$	7.8
Risk 3	10	$= \sqrt{(5^2 + 6^2 + 10^2)}$	12.7
Risk 4	8	$= \sqrt{(5^2 + 6^2 + 10^2 + 8^2)}$	15.0
Risk 5	5	$= \sqrt{(5^2 + 6^2 + 10^2 + 8^2 + 5^2)}$	15.8

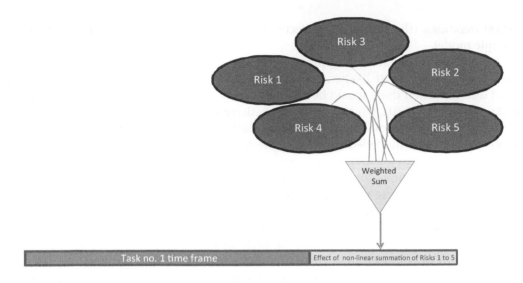

Figure 4.16 A weighted sum: a non-linear addition

In Table 4.22, five risks may affect a task. In the second column, the current sample of each is shown. This is the number of days extension the risk model determines will be required to absorb the risk on the current iteration. All five risks have finite durations, and so the data in the table shows a rare coincidence of all five unwanted outcomes materialising at once.

The maximum discrete extension of 10 days, Risk 3, was deemed too little to deal with a situation in which the unwanted outcomes of the other four risks occurred at the same time. Equally, the sum of all of the unwanted outcomes, 34 days, was felt to be too much. In the right-hand column is the square root of the sum, of the squares, of the unwanted outcome durations – Risk 1; Risks 1 and 2; Risks 1, 2 and 3 and so on. Reading down the column shows the extension in duration of the ensemble as each unwanted outcome piles on top of the previous one. This increases with each unwanted outcome, but not by its full value, giving an overall extension of just under 16 days for all 5.

There are other algorithms for weighted sums you may prefer to use, perhaps one that gives greater relative weight to successive unwanted outcomes that have occurred, and hence models the effect of an increasing load being placed on a finite set of resources available to contain them. Whichever algorithm is devised, avoid using a simple accumulation when it is not appropriate.

The Time Function

Running the 'risked-up' version of the time risk model in the same way as the cost model was run, will give us insights into the project plan in the same way as the results of the cost risk model did for the estimate.

The output is the time function, from which we can read the earliest completion, latest completion, most likely most likely completion, exposure to overrun and the opportunity for an early finish. Figure 4.17 shows it in probability density form after

10,000 iterations (the results are collected from cell E18 in the summary sheet of the example model.

The minimum duration is approximately 130 weeks, the maximum approximately 190 weeks and the most likely, approximately 155 weeks. If the latter were made the target duration then there would be an opportunity for the project to complete up to 25 weeks earlier (155–130). On the other hand, there is a risk that it may finish 35 weeks later (190–155). A vertical line at 137 weeks marks the duration of the underlying plan before the risk was modelled: this splits the function in two at a point that says the planned duration has a 99.5 per cent chance of being exceeded.

From the cumulative form below (Figure 4.18), we can read off the probability that the project will complete within a desired time frame. If the durations are converted to dates by ascribing a start date to week 0 in the model, a desired date can be read. These results are taken from cell E21 in the summary sheet of the example risk model.

With a little ingenuity it is possible to pick-off time functions for other milestones of interest and to compute all of them in a single modelling run. For example, cell U15 in the time model sheet is the start date of the piling works and is thus effectively the milestone for starting the on-site works. If I make this an output cell from the model in the summary sheet, see E24, I can calculate the range of finish dates for the immediately preceding phase of the project, which is effectively the design phase (Figure 4.19).

The interim milestones are very useful for reporting possible delays in the start of important project stages, for example:

- The range of completion dates for financial closure is probably the time function for the start of detailed design.

Or

- Delays to start of construction has the same time function as delays to the completion of design that usually precedes it.

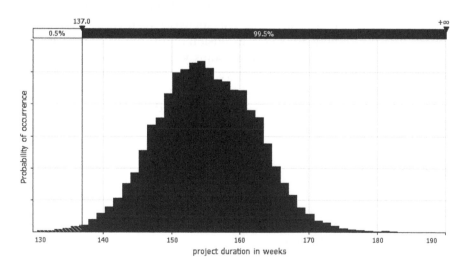

Figure 4.17 The time function for the example model

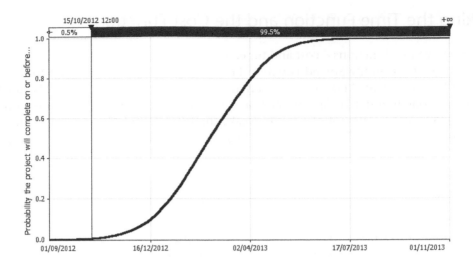

Figure 4.18 A cumulative time function in calendar form showing chance of completion on or before the planned date is 0.5%

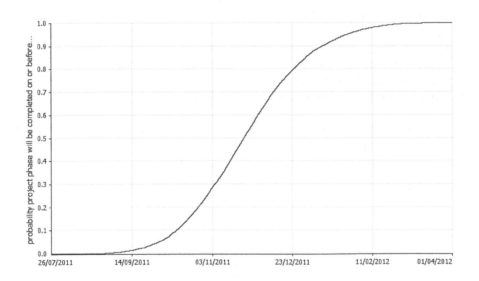

Figure 4.19 Range of finish dates for the end of the design phase

These are valuable insights into how the project may unfold and it would be a foolish manager who planned to let resources leave the project on their planned date without reference to the relevant time function. Given Figure 4.19, would you book ten yards of pink ribbon and royal personage to turn up on week 137, 18 October 2012? Neither would I. A year later would be more like it. And just as importantly, given Figure 4.19 and knowing that the design phase could run on until early April 2012, I would put an option in the contracts for key designer staff for them to be retained until then. This is good project management based on good risk analysis.

Linking the Time Function and the Cost Function

We now have two functions – cost and time.

In the section on the Spread Duration Primitive (see Primitive 7.1 duration spreads on page 113), I explained how the time function could be converted into a RMT function for use in modelling time risks. I wrote that if in a freak scenario, the time function had to be spread over a range of one to six weeks then the appropriate model of the time function would be a flat dice. But since it turns out to be spread over a different range and exhibits a peak, this was equivalent to a dice with more faces, biased towards the values in the area of the peak. With reference to Figure 4.20, the dice would have to have the maximum value of 180 minus the minimum value of 122, plus 1 which equals 69 faces, and be progressively loaded so that values around the peak of 155 occurred more often than those tending towards 180 or 122.

We can use this biased dice to generate the number of weeks the project could take to complete, and if we were to multiply that by a time-dependent cost such as the rental of office space, we would be modelling the range of costs of office space rent. This means that a time function can be used as a spread on quantity primitive for time-dependent costs, of which there could be many.

Indeed there can be several time functions in a project, perhaps one for each phase, with their own particular unit time dependent cost. These may include:

- Management fees – for the duration of the project.
- Late penalty payments – on the time after completion was due.
- Plant hire – during the construction phase.
- Office space rental – for the duration of the project.

And so on. Complex models will therefore often have a set of time functions driving the quantities of time dependent costs.

The connection of time functions to costs in a model can be done in two ways. In the example model, the cost part and the time part exist side by side. When the model is run

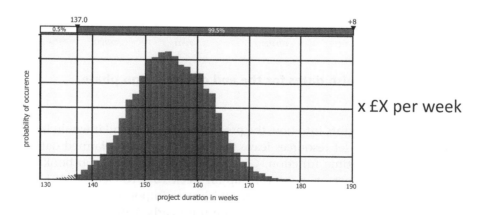

Figure 4.20 Using a time function as a spread on quantity primitive for time-dependent costs

they both iterate concurrently and I can therefore use the sample of time to drive directly the quantity of cost as is done in Risk 12, cells A81 to G81 of the cost model (Table 4.13):

Table 4.22 Using a time function as a spread on quantity primitive for time-dependent costs

Row/column	A	B	C	G
	No.	Synopsis	*Modelling note*	Sample
81	12	Project management charges are per accounting period for the duration of the project.	*Time-dependent: spread is based on overall duration of the project.*	= 'summary sheet'!B12*'time risk model'!T52

If you follow the references through in G12, summary sheet B12 is the project management cost, which is assuredly a time-dependent cost. This is because if the project overruns, as Figure 4.15 predicts it will, the marching army of the project team, myself among them, will still be filling in their timesheets on a Friday afternoon. Time risk model T52 is the percentage overrun on the planned duration of 137 weeks. The sample evaluated in G12 is thus the additional project management cost of an overrun time.

This direct method is simple to assimilate and encode. However, running the two models concurrently does dramatically increase the possibly already heavy computational load of either of them. While it may be impressive to run both cost and time models side by side, it looks less so when a minor change to one cost risk requires an embarrassingly long time to gauge the consequences. Therefore it is useful in practice to replace the time model by its equivalent function using either the curve fitting tool or the RMT Cumul function. This is the indirect method (and is a sensible thing to do) because although plans are twiddled and tweaked endlessly by their developers, those who need them for other purposes, such as project managers and investment analysts, tend to wait for the froth to subside and a definitive release to emerge. As a result, the emulations underpinning time models tend to be static, typically two or three months at a stretch, meaning there is an awful lot of computational load that can be avoided. This is especially useful in circumstances where risks are being revised day by day and the current exposure re-computed frequently.

Below is the time function with a curve fitted to it (Figure 4.21). The RMT formula for the curve is:

= RMT/betageneral (14.318, 15.762, 115.728, 192.55)

and I could use this in G12 of the cost model:

= unit cost of project management time * ((sample from BetaGeneral – 137 weeks)/137 weeks)

to generate the cost of the predicted overrun.

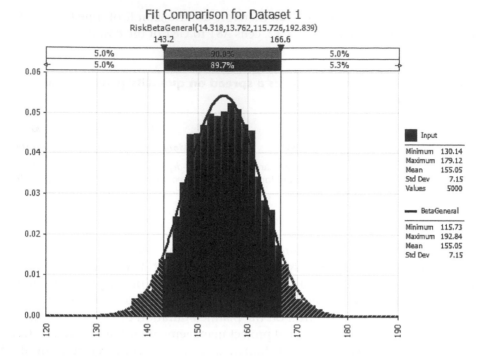

Figure 4.21 The time function for the example model with a curve fitted

Knowing that I will have to review and possibly revise the emulation and the mapping of the risks, I could then delete the time risk model from my working version of the full model whenever a new version of the underlying plan is released. This usually takes a few days to do.

Cross-linking time risk functions in any form – I casually refer to them as 'time spreaders' – and running both means that from iteration to iteration, changes in the duration of the project affect the costs of the project. The resulting cost function that emerges not only contains the effect of cost risks, but also the effect of time risks. I think this is a very valuable thing to know, and that it is the most powerful and illuminating result generated by risk modelling, so much so that I will start the next section by restating it in case anyone has decided to skip the end of this section.

The Integrated Cost and Time Risk Assessment: ICTRA

When I run the example risk model for 10,000 iterations I get the following results (Figures 4.22 and 4.23).

The figures show the cost function for the example project, its probability density and cumulative probability forms. Though the density curve is easier to assimilate visually, the cumulative curve is the easier one from which to extract useful information.

The minimum and maximum costs of the project are £15.25M and £18.5M respectively, both of which have a low occurrence probability. The most likely cost (the peak) is ~£16.6M. Though it has a higher probability of occurrence than either

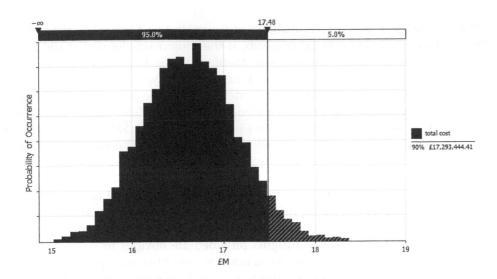

Figure 4.22 The cost function for the example project in density form

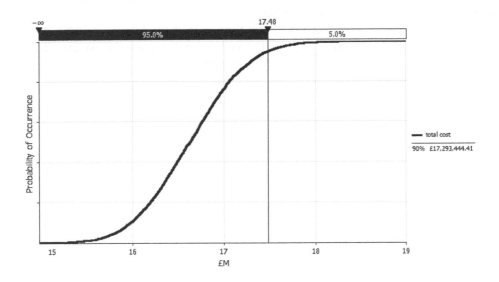

Figure 4.23 The cost function for the example project in cumulative probability form

the minimum or the maximum, surprisingly the most likely has a low probability of occurrence as well. This is because of the unspoken assumption that we mean *exactly* £16.6M, not £16.5999M or £16.0001M.

The statistic that makes more sense, especially to the project funders, is the probability that the project costs will be equal to or less than a particular sum. For this we need to refer to the cumulative curve. To illustrate, a vertical line to the right of the cumulative curve shows there is an 80 per cent probability that the project cost will be less than or equal to £17.07M.

A line to the right shows the probability of the project cost being less than or equal to £15.875M as 6.5 per cent. In other words, if the project was funded at the value of the estimate, i.e. without the risk exposure, there would be a 93.5 per cent probability it would be exceeded. Yet by using the cost function astutely, it is possible to set a fixed budget and set aside an appropriate contingency fund in case an overspend looms. For example, the project team could be given a budget of £16.5M from Figure 4.23, and the funders could set aside a contingency fund of £17.07M – £16.5M = £0.57M to make the total funding sufficient for it to be delivered on budget for an 80 per cent level of confidence. More money could be made available if the project effected others, or had a high public profile. The cost function generated by a combined cost and time risk model is a very valuable and informative result. It not only shows the cost of the project, but also the financial effect of both the risks to costs *and* the risks to time in a concise and clear way. The spin-off result of the time function is no less valuable and informative. The visual presentations mean the insights provided are obvious to see, interpret and challenge in a way a table of numbers could never match. They are a powerful aid to the negotiation of an agreement on project funding. Would anyone willingly accept a budget equal to the estimate if they had an analysis that show this has a 93.5 per cent chance of being overrun? Or, from the funders' standpoint, not make provision for an overspend of £3M in their funding forecast? No and No.

The combining of the cost risk model and the time risk model into a single assessment is a very powerful and illuminating tool. I call it an Integrated Cost and Time Risk Assessment, an ICTRA. Its rarity continues to surprise me.

5 *Using Risk Analysis to Inform the Allocation of Risk Ownership*

Background

In order to calculate the funding needed by a project it is important that the analyst takes an all-round view of the risks. They must take into account all the risks that are identifiable when researching and constructing a risk model, without pre-judging whether they will be held to account by their particular client. With only a few exceptions, it would be acceptable to base the modelling solely on those risks taken on, or to be taken on by that party. I would try to account for all the risks if only to reassure myself, and to show my client, that risks that are not theirs are somebody else's. There is nothing like a risk no one owns for causing dispute and conflict between stakeholders

A project in development always arrives at a stage when the risks have to be allocated between the parties to it. This implies the cost function has to be notionally partitioned between them, and so the question of risk ownership – who is exposed to what – arises.

To my mind, this process of allocating risk ownership sits on the cusp between risk analysis and risk management. Before it, a risk analyst can take an all-seeing view of the risks without overly undue consideration of what should be done about them. Researching and proposing treatments and ownerships should certainly not take up time that would be better devoted to researching and delivering a dependable, valid, funding analysis to the funders – within reason of course. Further, the risk manager will not bother the risk analyst because it will not yet be known if a risk is theirs to control, again within reason.

After allocation, however, which is usually written into the contracts that define the commercial context of the project, risk managers will know who has which risks, and the process of risk management can proceed apace. For the risk analyst to advise then that the funding calculations are not finished would be tantamount to negligence. In a perfect world, before contractual allocation, the analyst informs the funder, and after, informs the manager, as I will discuss later. By informing I do not mean handing over a list of risks that need to be dealt with, but instead the reorientation of the analyst's thinking to what the manager needs and how this can be done.

Advising on the allocation of risk is the last act of enlightened detachment the analyst is called upon to perform. It will help all parties if the allocation is a good one, so I will now explain how I go about it. My wish is that no manager should despair on receiving news of a risk they have to manage, but instead think it is one they can do something about. Though it has to be accepted that often managers are asked to manage risks they cannot mitigate (for example, shortages of expertise).

In what follows, I will generally assume there are just two stakeholders to the funding of a project – the funders (even though there are usually several investors in a project), and the project manager (there is usually only one of these). Under the project manager I have assumed there will be the familiar raft of consultants, contractors, project services staff and so on.

The Meaning of Risk Ownership

The traditional method for the allocation of ownership takes a managerialist stance. It is that risks should be owned by the person or persons best able to bear them. This is usually assessed by managers according to the adjudged expertise and experience of those who have some connection with the risk. This is fine but it is an approach predicated on the existence of an allocation of risks between the parties, to a project that has been contractually established. The managers will then be divvying up the risk management actions and responsibilities for risks which the party they represent has had allocated to it by the contracts and agreements that constitute the commercial framework for the project. It is essentially a post-contract activity.

The role of the analyst in allocating risk precedes this. It is to advise what the allocation to the parties to the deal should be before those contracts are signed, i.e. pre-contract.

The distinction between allocating risk pre-contract and post-contract is important because risks cannot be allocated contrary to contract or to the law. The approach I recommend for allocation of ownership, pre-contract, is based on an understanding of the costs associated with a risk.

1. The cost of prevention. This is cost of trying to stop the unwanted outcome of risk materialising. For example, if the risk is a fire, the prevention costs could be the costs of signs banning naked flames.
2. The cost of containment. This is the cost of dealing with unwanted outcome. This tends to be incurred during the period, usually sudden and short, in which the unwanted outcome materialises. For example, putting a fire out, removing goods and equipment, transferring operations and such like.
3. The cost of remediation. This is the cost of restoring the works of the project to the state they were in before the unwanted outcome occurred so that the project can resume.

The first, the cost of prevention, can be either a matter of ordinary business, in which case it raises no additional costs, or extraordinary business, in which case it does. An example of the former would be:

There is a risk that survey teams will not be able to gain access to the site.

The cost of prevention here would be the time of the project manager arranging the access, something I would consider a matter of ordinary business, thus incurring no extra cost. On the other hand:

There is a risk that protestors will attempt to prevent the surveys proceeding.

This may well require the introduction of special expertise and procedures (for example: security, consultation exercises, weekend working...) which I would think could be a matter of extraordinary business. The project manager may not have budgeted for these prevention costs because the need for them has presumably only emerged with the identification of this risk. Project managers tend not to be given budgets and resources for things that are not certain to happen in case they are tempted to divert them to other things. This is simply a matter of traditional cost-control practice.

The costs of containment are almost always extraordinary for the same reason. To extend the survey theme, the discovery of asbestos linings would be highly likely to incur extraordinary costs of containment because the estimators would not have estimated for such a discovery.

Further, the subsequent removal of any remaining asbestos in order to restore the work site to a safe environment would be another extraordinary cost of remediation.

Here is the nub of it – when recommending the ownership of a risk, it should be allocated to a party. By doing this, an analyst is saying that the party should accept liability for these three costs. This is what ownership of a risk means; the owner pays the costs of prevention, containment and remediation. This is what should be made clear in the contractual framework of the project.

Digressions: A Manager has to Deliver and an Analyst has to Decide

It is the job of an analyst to decide. It is an article of faith among managers, and project managers especially, that they must deliver. They achieve this not solely through their own efforts but by arranging for it to happen within a team. They can be said to facilitate the delivery.

As the manager must deliver, so must the analyst. They must facilitate the decision, which is to say that through coordination of the thinking of others they must bring a consensus into being. This consensus need not be unanimous, but it should be one that is practical, tolerable and fit for purpose.

Some, perhaps many, would say an analyst does not decide, but rather has to recommend and then let others decide. While I don't disagree with this, I do think it is an analyst's responsibility to research risk diligently and thoroughly, to model it correctly and completely and to interpret their results thoughtfully and with integrity so that they become convinced of the right decision. Analysts must set about their work as if they were going to make the decisions that will be based on it, even if at the last moment they pass the ball to someone else.

The two most common decisions an analyst must make, or rather inform, are the quantifications of the risks and their ownership. Perhaps the most efficient way to achieve these is to decide what the answers should be and to offer them first for consideration and then for confirmation or correction. There are three reasons for this:

First – analysts, particularly experienced ones, often know as much about the nature and content of the decision as the person they would turn to for an answer. Not to offer their own opinion when seeking that of their interlocutor would diminish the debate and so possibly prejudice the robustness of the answer. Two heads are better than one.

Second – I find it quicker to reach a consensus if I offer an answer for consideration. I recommend this is done in person by telephone or during a visit, and not by despatch of an e-mailed form that has to be filled in and returned by the recipient. No one I know likes to receive forms and any analyst who resorts to such an impersonal approach deserves all the late and inadequate responses they receive. They shouldn't be surprised when support for a decision they have negotiated crumbles. Humans are designed to meet and to discuss, and are given several ways to stimulate an open and cooperative discourse; body language, tone of voice, choice of words, stress on syllables, nuance of argument and so on. Not to use these to achieve a consensus on risk matters, and indeed any other intellectual position a project needs to establish, is foolish. Analysts must decide what would be a good answer to the decision and then discuss it. An agreement on what is then the best answer will be arrived at faster.

Third – unlike the 'complete-the-form-and-send-it-back-to-me' researcher, it is likely that the analyst will need to write up the agreed answer and the reasoning behind it for the record, and what we write we remember. This increases an analyst's knowledge for the future. What others write, we forget.

Further Risks to Take Into Account

Risks are usually researched within a community of designers, constructors and the operators of the asset the project is intended to deliver. This can produce a bias in the contents of the risk register towards things that could go wrong, or that cannot be done in respect of the works: shortages of resources, material price fluctuation, things breaking, delays in approvals and so on. Since these risks tend to be identified from research into the works of the project, I call these the Works Risks. However, before the analyst starts to analyse what the ownership of these risks ought to be, it is prudent to step back and consider some other areas of possible risk in case not everything has been discovered.

The first and most fertile area to research for further risks is the commercial framework – the set of contracts and agreements that govern the project. Here is a trivial illustration of why. Say I had a pencil with a broken point and you had a sharpener. If I ask you to sharpen the pencil for me, you would probably think this is a low-risk thing to do and get on with it. But if I said that if you break the point you must pay me £5, you would probably decline to do the job. What was a low-risk job has become a high-risk job, not because of the work itself but because of the contract I offered you. It is the same with projects. Contracts can introduce more risks, and since these risks relate to the transactions for the project, I call them the Transactional Risks.

Here are some real examples of risks that did not appear in my researches into works risks on a project, but did when a colleague researched the transactional risks in the commercial framework:

1. A penalty of £1000 per hour or part thereof will be payable for a failure to return the construction site to operational use by Monday 0600 after weekend working. (Modelled as a time-dependent cost to the contractor and as a matching income stream to the project manager)

2. Party A retains the risk of additional cost or delay if B cannot obtain the necessary consents. (Modelled as emerging cost and delay similar to the landslips risk in the example model)
3. The form of contract is not defined and so it may be one that does not permit the obligations to flow down from this agreement leading to Party A retaining risk that would otherwise flow through the supply chain. (Modelled as a strategic risk – see strategic risks.)

And so on..

It is therefore prudent for an analyst to analyse the commercial framework *before* starting to analyse the allocation of ownership.

The contracts the analyst should research are those that have been signed and those that are being drafted, that is to say the definite and the probable. They fall into two groups; those between the funders and the project manager for funding, and those between the project manager and the contractors for the provision of goods and services. The former are sometimes called the funding agreements, but they are contracts just the same (Figure 5.1).

A project management team is often less aware of the obligations set out in the funding agreements between the funder and the project manager, than it is of the obligations set out in the contracts for others to do the work. This is because these agreements tend to be much more confidential, secret even, and so are seen less often. Unlike the project manager–contractor contracts, which tend to be based on tried and tested (in law) industry-standard templates, the funding agreements are bespoke and so past experience is of limited help, and whereas both sides to industry standard contracts tend to understand their rights and obligations, this cannot be consistently said of funding agreements.

The analyst must study them however in order to decide what the ownership of risk between the funder and project manager is, and must also study the standard forms to decide what the ownership should be for the project manager and contractor. Thankfully, in my own experience, both the agreements and the contracts are usually clear and concise about who owns which risks (they are usually written down as the costs a party shall bear) because there is no reason for them to be otherwise.

Since project managers and contractors are very familiar with the standard forms of contract, there is always plenty of advice and opinion to be had on what risks they contain. All the analyst has to do is ask. The agreements, however, tend to be created anew for each funding stream of each project. Their detail tends not to be known by the project managers as comprehensively as the contract detail because, with reference to Figure 5.1, each party's attention tends to be oriented to the right. Generally therefore,

Figure 5.1   The contracts flow

the analyst must read the agreements. In my experience they are generally never as long as the contracts are, and they will have not have the complex phrasing intended to avoid disputes and claims that contracts have accumulated from being tested in court.

There are four other good reasons for researching transactional risks.

First – if some of the contracts have been signed, then it is important to know how these contracts have allocated ownership of any risks they cite. The analyst cannot decide on a risk ownership that is contrary to contract or to any applicable legislation. If a contract states:

The consequential costs of a failure to relocate utility services crossing the side are the contractor's.

Then allocating ownership of a risk such as:

There may be a delay in diverting the gas pipeline crossing the site

to the project management company on a risk form in a risk register is wrong because the project manager will not have to meet these costs if they fall due, the contractor will. The contracted positions on ownership have precedence, and the analyst needs to know how ownership may already have been decided, lest they discredit their own analysis of it.

Secondly – the analyst may decide there is a compelling case for a risk to be owned by Party A only to discover it has been contracted to Party B. In these circumstances, this misallocation in itself could be a risk (see example 2 above in which it would be better if Party B was liable for the consequences of failing to obtain the consents).

Thirdly – the analyst must think holistically about the transactional risks because there may be possible financial liabilities about which the contracts are silent. For example, if there is a risk of delay diverting a gas pipeline, the project management company may have to bear the costs of all the other contractors affected by this if the contract says nothing about passing these on to the company diverting the pipeline. See Example 3 on page 163, where an agreement does not define any form of contract that would allow its terms and conditions to be passed along a supply chain. The agreement is silent on this important matter, and that silence is a risk in itself.

Fourthly – the analyst may be able to affect some early days risk management work by proposing changes to agreements and contracts that would eliminate risk, but have not yet been signed. This may be possible to effect even in agreements and contracts that have been signed, if these contain provisions for the alteration of their terms and conditions. Indeed I think it is imperative that the analyst does so. If this opportunity is missed, there is no legal obligation on anybody identified as a risk owner in a risk register to act to manage their risk. Risk management would then be reliant on altruistic good sense, benevolent goodwill or wishful thinking, none of which will provide cast iron assurance to the funders that risk management will happen. This is also a risk in itself.

Transactional risks are not always explicit statements of which party will bear the cost of which risk. They can emerge in the gaps and overlaps of the commercial framework. Contradictions, for example, are not uncommon and risks can sometimes be discovered in them. For example, a funder–project manager agreement I know states:

The system will be introduced into service on 1 September 2015.

But the project manager–contractor contract contradicts this:

Trials shall commence on 1 September 2015.

Here there is a misalignment of what the funder and project manager expects to happen on 1 September. There is a transactional risk that the project introduction into service date may be delayed by the length of the trials period.

Here is another:

Funder–project manager agreement:

The project manager shall not hire the services of legal representation. These shall be provided directly by the funder.

And:

Project manage–contractor contract:

The project manager's client shall pay for the services of the contractor's legal representation in the event a public inquiry is called.

This is a transactional risk to the project manager, because they may have to pay for legal representation for the contractor without being able to claim it back from the funder.

Some transactional risks will however be duplicates of works risks, though this may not be explicit when the wording is different. For example:

The project manager shall bear the costs of the removal of buried objects.

(In the transactional risk register, and in a signed agreement.)

And:

There may be old foundations in the area from the demolished gas works

(In the works risk register.)

Though it is right that the entry in the works risk register should remain in place so that the risk can be managed, it is the transactional form that has precedent because being written in a contract, its ownership is specified. It would be wrong if the risk appeared twice in the cost function because it appeared in both risk registers.

To conclude, the analyst must research the detail of the commercial framework if their work is to be complete. I would call into question any cost function or risk register I was asked to review that contained no evidence of transactional risks having been researched. Transactionals are an important area of risk exposure, and like a cost curve without a time risk modelling component, a cost curve without a transactional risk component is probably invalid.

I know of a project that has a 95 per cent risk exposure of ~£3B, of which ~£700M is transactional risk arising solely from the contracts and agreements. That is a significant quantum of risk to leave out.

Risks in the Assumptions

Below is an extract from a list of assumptions made by a team of designers on a railway project to allow them to progress their work (Table 5.1).

Table 5.1   Team A assumptions

Number	Assumptions from Team A
01	24 trains per hour – the planned opening peak service is 24 trains per hour in each direction and this pattern occurs at weekday morning and evening peaks
02	Fleet size – the overall fleet size is 63, 10 car trains
03	Reserve in fleet size – approximately 10% of the initial fleet size requirement (i.e. 6.5, 10 car trains) is identified as reserve rolling stock
04	Rolling stock specification – the rolling stock specification is based upon a 25kv AC overhead single voltage system (with space provision for dual voltage equipment) and a ratio of 2 powered units to 3 non-powered units across the fleet
05	Capacity: it is assumed that existing supply capacity for material can meet the project's demand
06	Franchise – there will be allowance in the costs for tendering the operating franchise

01 to 04 contain precise design parameters, and 06 a declaration of intention to do something. This 'something' should be reflected somewhere in the transactions for the project.

These assumptions will probably have been made early on in the project life cycle, and in isolation, in order to facilitate starting the design work. Isolation is probably too extreme a term, but the assumptions are likely to have been made with only limited consultation. Any wider consultation would have been reliant on others obtaining a copy of the list and reading it.

There may be risks to these assumptions being sustained. Other teams on the same project may decide upon different design parameters, or some assumptions will not have reflected in the commercial frame work being drafted elsewhere (Table 5.2):

Table 5.2   Team B assumptions

Number	Assumptions from Team B
01	20 trains per hour
06	Franchise: a design competition for the operating franchise will not be part of the delivered works

The conflicts are potential risks that the analyst must include in their analysis, and to find them will require time and effort to collate the assumptions made by different teams, and to align these so that the conflicts become apparent (Table 5.3).

Table 5.3   Mapped assumptions

Number	Assumptions from Team A	Assumptions from Team B
01	24 trains per hour – the planned opening peak service is 24 trains per hour in each direction and this pattern occurs at weekday morning and evening peaks	20 trains per hour
02	Fleet size – the overall fleet size is 63, 10 car trains	
03	Reserve in fleet size – approximately 10% of the initial fleet size requirement (i.e. 6.5, 10 car trains) is identified as reserve rolling stock	
04	Rolling stock specification – the rolling stock specification is based upon a 25kv AC overhead single voltage system (with space provision for dual voltage equipment) and a ratio of 2 powered units to 3 non-powered units across the fleet	
05	Capacity: it is assumed that existing supply capacity for material can meet the project's demand	
06	Franchise – there will be allowance in the costs for tendering the operating franchise	Franchise: a design competition for the operating franchise will not be part of the delivered works

Some assumptions, like 05, Capacity, may simply not be realistic (Table 5.4).

Table 5.4   An unsustainable assumption

Number	Assumptions from Team A	Assumptions from Team B
06	Capacity: it is assumed that existing supply capacity for expertise can meet the project's demand	This is unlikely because two other projects are currently absorbing all of the available expertise and the introduction of this project, a third, is likely to necessitate another source (training, overseas etc)

I would write up 02 in the above table as an unresolved design option, 05 as an emerging cost and 06 as an emerging time delay. I would then model them accordingly.

As with transactional risks, I would question the validity of any cost function that showed no evidence of conflicting or unsustainable assumptions having been considered in its derivation. The production of a coherent and cross-checked register of assumptions as a by-product of risk analysis, just like a logic linked plan, is often appreciated by clients and their project teams.

Strategic Risks

As discussed earlier, the work of identifying strategic risk and its subsequent management and reporting seems to have become a specialty in its own right, arguably outside the scope of this book. However, I do think that a risk analyst working among the muck and bullets of cost functions and risk management plans can and should bring something to a strategic risk forum.

The normal approach to identification of strategic risk is for a risk identification workshop to be held at board level in a company, or at least with a quorum of highly placed people. The role of the analyst is to ask why the board may not achieve the objectives of its business plan and to note the answers. The workshop may reveal many issues and knowledge gaps, the airing of which may consequently stimulate mutual support and coordinated action among the attendees. This alone would make the time well spent, but the analyst will be a scribe tasked to record the proceedings, and afterwards to make sure there is be good evidence in the company archives to demonstrate that proper governance in respect of risk is in place.[1]

In my experience, a board will be able to identify and describe strategic risks, and explain what is being done about them, without much being required in the way of facilitation. It will, however, expect those risks to be incontrovertible and it would certainly be a brave analyst who challenged them or sought to add to them. However, I strongly suggest that the analyst should pluck up their courage and pitch anything pertinent from their own work into the discussion. It would be canny of them to add an item to the agenda called 'Strategic risks identified by the analyst' in order to secure a right to speak.

---

1   See Footnote 2 on page 60, on corporate governance for a reference as to why this is the case.

There are seven areas an analyst could report. The first is risks that are being managed elsewhere and so do not need action at board level. A board may even appreciate being made aware of this because it would free some of their efforts to other matters that are in need of their attention.

The next four things to report should be researched at the same time, as all the other risks are not taken to be an additional task tagged onto the end of a risk study. I keep an area of my notebook reserved for any strategic risks I identify, in anticipation of being asked to attend a strategic risk review at some stage during a study.

Secondly – note and report those functional areas of their business that seem to have insufficient capacity to process the work required of them, for example:

- Documents that are being held up in a procurement process may indicate a problem with the Procurement Department, or with the quality of the documents it receives.
- Maintenance and repair time that is being lost or extended because staff do not turn up, or by tools and materials not being available.
- Complaints heard about a shortage of resources, which may indicate a problem with a Human Resources Department, or a shortage of expertise.

Thirdly – an analyst should note and report risks that are the consequences of parties over which the project manager has no control. These may be quite low level, below the parapet of a board, but even so a strategic risk review is a good opportunity to air them. For example:

- Another project may not complete works on which the project in question is dependent. Perhaps it needs to use the same site or the same work force.
- Parties may be found to be of a mind to exert their legal rights to disrupt or delay the project in some way.
- Parties insisting that the project must be executed in a less efficient way than the planned disruptive way – say in the sequencing of its construction works.
- Parties delaying authorisation of paperwork to permit access to, or exit from worksites.
- Parties refusing to accept back into their own operation and maintenance the assets that the project has had to take into custody for protection, refurbishing, decoration etc. while the works are under way.
- Parties wanting the project to provide works, typically amenities, that are outside of its scope.

Fourthly – the analyst should report risks that cannot be modelled. Risk modelling is always done on a presumption that given the right amount of time and money, a project can be done (but that it needs both a cost estimate, a plan and a risk assessment to find out what these are). However, there may be risks whose unwanted outcome may prevent completion no matter how much time and money is made available. These have not been possible to model in the cost function. I was taught that if something has to be hidden in a report, the best way to conceal the information is to write it into a table – no one reads them. Since these unmodelled risks may go unnoticed if they are reported in a tabulated risk register, it is good practice to enunciate them in a meeting because they can be significant. For example:

- A change in government may lead to the halting of a project, or an alteration of the terms under which it has been authorised.
- Procedures and processes are being enacted that do not conform to current or emerging legislation.
- The demand of the project for labour, machines, materials etc. exceeds the available supply.
- The project cannot be delivered for the funding available.
- The project cannot be delivered within the mandated timescales.

It is probable that no contingency sum would compensate for the unwanted outcomes of risks such as these because they could very likely lead to the project being stopped. So whereas risks like these should be reported and incorporated into a risk management process, I would not expect to see sums of money and time for them in a cost function. Indeed the idea of a sum of money and a period of time being set aside for a project that would be stopped is spurious unless it is for closing down and tidying up.

Fifthly – the analyst should report small risks whose combined effect is significant. These are risks that crop up time and time again across the project and individually suggest little is wrong, but collectively tell a different story. For example, say a construction company has ten work sites to start and only nine tower cranes. At a worksite level, each has a 90 per cent chance of getting a crane. But across all ten sites, one of them will not get a tower crane. This might be significant at the portfolio level of all the projects. Risks like this are common, where demand just exceeds supply, be that for equipment or labour.

Sixth – a key characteristic of the modellable risks is that they are probably strategic, that is to say their unwanted outcomes may prevent a business achieving its declared objectives. Something that should make the analysts knees knock is that their treatment needs to be in the hands of the executive because it is beyond the powers delegated to the project manager to effect their reduction or elimination. If the analyst will not stand up for the project manager in such a forum, who will? Perhaps no one. Take courage and speak up.

There is a final reason for the analyst to speak, and it is a tough one. The analyst, when writing up synopses of what they think (or has been told), are the strategic risks, should map them to the client's organisation chart using the traditional criterion that the party best able to deal with the risk, owns it. If the analyst then finds there are some risks whose treatment does not naturally sit within the remit of any of the executive members, arguably that board is not structured to deal with the risks it has. This is perhaps the most important strategic risk of all. Indeed there may even be board members whose areas of responsibility are found to be entirely risk-free, and it will take a strong analyst, convinced of the correctness of their mapping analysis, to stand in front of them and say that the board has the wrong mix of expertise on it. It should be said that if the board cannot take the risk management actions necessary to secure funders' investments, then questions must be asked as to whether certain members of the board should be there at all.

Deciding Ownership of the Remaining Risks and Discovering more Risks: Interface Violations and Funding Blocks

Having studied the contracts and agreements to identify transactional risks, the ownership of which are explicitly, contractually determined, the analyst now has to decide who owns those risks that

- Have not yet been assigned to a party to a contract.
- Or have been assigned to a party to a contract but now need to be assigned among the party members.
- Or have been assigned to a party member who is considering transfer of the risks to other parties, e.g a subcontractor, and is in need of advice about which ones, to whom and why so that their choice is informed and, if called into question, rational and justified.

Though these three scenarios are different, they all benefit from the same analytical approach. In the first case the analyst will want to provide an insight that will help the legal teams to assign ownership effectively (those risks will of course then become transactional). In the second, seeking a similar insight that would allow the project manager to delegate responsibility for managing the risks to their team members with the intention of setting a process for periodically reviewing how well they are doing on these tasks and reacting accordingly, a role sometimes delegated in its entirety to a risk manager. And furthermore, as I will show, identifying those risks that have been made the manager's responsibility, but for which they do not have the authority or the means to act, a conundrum to which the third scenario may be the better solution.

The usual technique for deciding risk ownership is to allocate it to the individual, team, department, subcontractors etc. best able to bear it. This can so often generate a load of nonsense that will not withstand the rigours of project delivery because, just like among my family I am best able to bear the mortgage on this house, it does not mean I actually can. When an analyst recommends ownership of risk, they are providing an implicit assurance to the funders that that risk can be managed effectively. Too often one sees risk registers in which ownerships of many risks are escalated to a senior executive, as if a projects director is ever going to study a risk register in order to plan their day according to its injunctions, or to a junior staff member who has no empowerment to act. An analyst must be realistic and practical in deciding ownership because they have a duty to see the risk exposure reduced, even if only by stimulating the actions of others.

To assign ownership to one who is not likely to, or cannot, act says to me that completing the form is more important to the analyst than reducing the risk. I am tempted to ask analysts whose allocations of ownership persistently lean towards the high and mighty to walk the upstairs corridors and politely knock on the solid doors asking for news of progress. If an analyst is not prepared to ask or help the person if they can or are prepared to reduce or eliminate a risk, then the analyst must question the quality of their risk allocation choice, keeping in mind their duty to the investment the funders have presumably made.

Allocation of ownership requires thought and negotiation skill. It is akin to a strategy game about how to outsmart risks that are out to get you. I do not want to tend towards anthropomorphism here but risk is the enemy of projects, and intellectual rigour on the strategy and tactics for defence against it is entirely appropriate.

Before I explain the analysis of risk ownership, I want to distinguish between the owner of a risk and the owner of an action to manage the risk. The first is nearly always an organisation because only they, or individuals trading as such, are prepared to accept the legal liabilities for risks. The organisation's owners sometimes hedge their risk through insurance, or by only having a small percentage of their funds invested in a single project so that if it fails their losses are bearable. On the other hand, the owners of actions intended to reduce risk are usually named individuals within the organisation that owns the risk, whom auditors and reviewers can then contact to see if the risk exposure position they are supposedly managing has altered.

In the analysis that follows, it is the former: the organisation as owner that I will be trying to determine. Who will then be its actionee is for it to decide afterwards.

The analysis begins with the three costs associated with a risk, which are, to repeat them:

1. The cost of preventing the risk occurring – prevention (or, alternatively, the cost of avoiding it)
2. The cost of stabilising the situation should the prevention strategy not be successful i.e. the unwanted outcome happens – containment.
3. The cost of restoring the project to the state it was in before the unwanted outcome of the risk occurred – remediation.

In terms of a risk of fire, the costs could be:

1. to try to prevent one happening
2. to put it out if it does
3. to repair the damage afterwards.

Or in terms of a risk of failure to integrate systems:

1. A systems integration plan that includes a complete set of common specifications, and a complete list of applicable standards.
2. An option to extend the tenure of key experts within the systems integration term.
3. The replacement of failing and failed components.

In deciding that an organisation is the one that should own the risk, the analyst is saying that it should be liable for each of these costs.

These are also three management issues to consider:

1. Who has the expertise needed to overcome the risk?
2. Who has the resources needed to overcome the risk?
3. Who has the obligation to overcome the risk?

Expertise and resources differ. The expert may be a single individual or a small company which, for their expertise to be effective, needs others to provide the resources – men, machines, materials, access, time, licenses, consents, money and whatever else may be needed. A computer programmer is a good example of an expert who is unlikely to have the development systems that will be needed to put their expertise into practice.

Obligation is different from either, and is the party that in the mind of the public or political community is expected to act to resolve undesirable outcomes. A good example would be BP and the Deepwater Horizon oil rig that exploded and sank causing the huge oil spill in the Gulf of Mexico. Precise responsibility for this accident does not matter to the public, nor to the politicians, because the obligation was placed on BP by the US Government to do something about it.

The costs and management issues form a 3 × 3 matrix (Figure 5.2):

	Prevention	Containment	Remediation
Expertise?	Party name?	Party name?	Party name?
Resources?	Party name?	Party name?	Party name?
Obligation?	Party name?	Party name?	Party name?

Figure 5.2   The risk ownership matrix

One such matrix should be created for each risk. The analyst has to then decide which organisation has the wherewithal; expertise, resources or obligation to carry out the prevention, containment and remediation tasks specific to the risk.

The entry should be the smallest credible unit within the organisation, for example an individual, team or department, not just the international PLC that may be their employer. The absence of such detail may obfuscate the reason for the organisation being identified, though of course for major risks the smallest credible unit may be the PLC itself. For example:

Containment/expertise: Analogue Solutions Team, Signalling Design Office, Reading UK – Global Signalling Solutions Inc.

This will probably help those being informed by the risk work understand why Global Signalling Solutions *should* own this aspect of the risk.

Containment/expertise: Global Signalling Solutions Inc.

Completing matrices is difficult and time-consuming. I have often felt the temptation, when faced with having to publish a rather sparsely populated set of them, to add in a few recommendations for ownership that I have not discussed with the potential owners. There is nothing wrong with this, and it may even be a good thing, the legal teams preferring discussions about who is liable for what taking place within their purview.

It is important to let those who use the ownership analysis clearly know whether or not the allocations of ownership are the analyst's recommendations alone, or if they have been endorsed in some way by the putative owners (Figure 5.3).

	Prevention	Containment	Remediation
Expertise?		Analogue Solutions Team, Signalling Design Office, Reading UK - Global Signalling Solutions Inc. Ownership discussed with GSS and provisionally accepted subject to agreement in contract.	
Resources?			
Obligation?			Suggested: National Rail PLC. No discussions have been held with National on this allocation

Figure 5.3   A part-completed matrix

Having completed the matrix for a risk, a question needs to be asked: is there a contract, or is one that obliges the three parties to act together to affect prevention, containment or remediation intended? If the parties were different departments from the same organisation then their obligation to act jointly can reasonably be assumed.

Further, is there a contract or is one intended between the organisation working on prevention, and the teams presumably standing by in some sense to contain or to remedy the unwanted outcome?

If there is no contractual obligation between these parties, either to each other or to the project that will hire them, then there are interface violations in the risk ownership matrix. This means the key objective of allocating risk ownership – that someone within the project will be doing something to reduce or eliminate risk – is itself at risk. Interface violations in the ownership of a risk are risks that should be added to the risk register, reported and if appropriate, modelled.

I hope I have shown that risk ownership requires a more thoughtful and astute analysis than putting a single name in a box on a risk description form might suggest. It may be suitable for identifying a risk reduction actionee but is inadequate for recording a proper analysis of ownership, a matter I think warrants a paragraph or two of reporting.

Some examples may help.

A new bypass has to go under a railway line. The solution is to excavate two large holes 10m × 10m square and 8m deep, one each side of the railway line. In one, a hollow 6m square section tube will be built from reinforced concrete. This construction is a short tunnel that will eventually pass under the railway. The plan is to suspend the railway services one weekend and during this time dig out the ground under the railway so that the two holes are connected and look like, from above, a grave with a length of Hornby-0 gauge stretched across it (though in practice the section of the rail affected will be cut out and placed to one side ready for reinstatement later). Then the concrete tunnel will be pushed by hydraulic rams from one box until it sits under the railway line. The space between the top of the box and the sleepers will be filled with suitable material to give the railway a foundation, and ramps will be constructed up to ground level from the ends of each hole either side of the railway. The entire construction – ramp, hole, box under the railway, hole and ramp will be further developed in a second stage of construction to carry the new bypass under the railway. This new under-bridge will have been constructed with the minimum of disruption to the rail services.

However, the ground is impermeable clay so there is a risk that the excavations may fill up with rainwater while the box section is being fabricated in one of them, or while the hydraulic rams are being set up, or while the contractors are waiting for the railway to be closed. The sequence of these construction stages will take a minimum of 26 weeks, a period of time when rain at some point is surely inevitable.

A follow-up line of inquiry an analyst may like pursue when looking for further risk (perhaps conceiving of an emerging cost chain model) would be to ask where does the pumped water then go, since water in the area presumably drains through the ground and so possibly back into the excavations? Foregoing this, the unwanted outcome is that the holes fill with water.

- The prevention action is to install pumps in the excavations to drain them whenever they need to be.

If this fails because the pumps do not have sufficient capacity, or because they break down, or in the actual case I am thinking of, the pumping subcontractor had not been paid by the main works contractor and so removed them, then:

- The containment action will be to stop the excavations filling up further and to pump them out.
- The remediation action will be to repair flood damage to the hydraulic rams, to the concrete of the box, and to the fabric of the side walls and bases of the excavations.

The ownership matrix is illustrated in Figure 5.4.

Reading from left to right, top to bottom, the pumping subcontractor will know what size and type of pump will be needed. On an assumption that a flooded excavation will have damaged the hydraulic rams and the concrete box they are used to push, the box push contractor will be the organisation with the expertise for remediation.

Re: resources: the main contractor will need to provide access to the site, work space, cranes, somewhere to dispose of the pumped-out water (with permission to do so) and probably labour, and so other than the pumps themselves, the main contractor will be the organisation with the resources needed for prevention, containment and remediation. I

	Prevention	Containment	Remediation
Expertise?	The pumping contractor *through contract with main works contractor*	The pumping contractor *not contracted*	The box-push contractor *not contracted*
Resources?	The main works contractor: labour, access, cranes, permissions to dispose the pumped out water *Through contract with CityBypassCo*	The main works contractor: labour, access, cranes, permissions to dispose the pumped out water *Through contract with CityBypassCo*	The main works contractor: labour, access, cranes, permissions to dispose the pumped out water *Through contract with CityBypassCo*
Obligation?	*Suggested: CityBypassCo* *No discussions have been held on this allocation*	*Suggested: CityBypassCo* *No discussions have been held on this allocation*	*Suggested: CityBypassCo* *No discussions have been held on this allocation*

Figure 5.4   The completed risk ownership matrix

accept the pumping contractor will have some resources too but for simplicity I will ignore these.

Re: obligation: I think that in the minds of the public and the press, the obligation to prevent the risk of the box filling with water, containing the damage, will remain the client's and the promoters of the bypass. This may seem odd, what do promoters know about pumps? But consider the 2010 Horizon Deepwater oil rig explosion where it is clear to me that BP has picked up the obligation to contain the problem because in the minds of the people, it more than any other party could have prevented such a thing happening. Here, any complaint from the public about delays in completion of the bypass will not be directed to the bottom of the contractual pile, but to the promoter sitting at the top. This would be on the grounds that it was in the best place to ensure the necessary preventions were specified in the contracts it awarded to the others. In the real project behind this example it was straightforward to say where the court of public opinion would judge the obligation to lie. It was easy to foresee a rising tide of public and political pressure to get the bypass opened.

I could accept an argument that the obligation for containment is really the main contractor's. In all probability the liability would be found to be the contractor's if a dispute over the costs of the flooded excavations came to court. However, such disputes can take years to resolve and should the press discover that the flooding has caused delay, it would be the client that stands in front the cameras.

Imagine the difficulty in managing an unwanted outcome if it is then discovered there is no contractual requirement to provide these resources to the people with the expertise. The party with the obligation to the public is usually the largest and most powerful party on the project, the client organisation, and usually it is providing the funding either from its own resources or as an agent for other investors. It writes the cheques at the top of the payments waterfall and so it has the power to ensure the contractual obligations are in place between the parties who may have a role in the prevention, containment and remediation of the risks. I suggest that it is nigh on impossible for a client organisation to pass the buck down the line when things happen that noticeably alter the expected outcome of a project. A client cannot avoid the obligation to ensure that risk prevention, containment and remediation can take place. Though there may be no obligation for them to pay for it because of the commercial arrangements between the parties to the project.

In some circumstances, completion of the matrix is easy because the expertise and resources are held by the same party. In others, the contractual relationships need to already exist. However, in circumstances where no cohesive action is likely because the relationships do not exist, this is risk in its own right that should be incorporated into the risk model, most often as 'not modelled'.

Deciding which party will be named in which cell in the ownership matrix can be an unnerving job, one that is far removed from the comfort blanket of many analysts. People will object to the choice and refute the reasoning that led to it but allocation of ownership must be done, and where it is permissible within acceptable limits of commercial sensitivity, acceptance by the parties named should be obtained. If this cannot be for any reason then this should be explained within the contents of the matrix. It is not an easy task, but then to be an analyst is to decide, or at least to bring, the analysis to a point where the answer as to who should own a risk is clear and compelling, and to then let others commit.

Interface Violation Risk

If there is no contractual obligation between the pumping subcontractor, main contractor and client for the purposes of prevention and containment, then such important actions are at risk of not being carried out. The pumping contractor and the main contractor need to be made aware of their obligations to act in these regards. This needs to be reflected either in the contracts between them or those between them and the client. If no such obligation is established, why then should they act to prevent or contain the risk? If this situation exists then that is a risk in itself. These fault lines at the interfaces within the ownership matrix I call Interface Violation risks. Here is an example:

> The box push is being set up below ground level in an excavation that is just larger than the box itself. There is a risk that the push will not go to plan and will have to be reset and tried again at a later date. Though the box-push experts are contracted to do this, the context of the work means they will need the assistance of the main works contractor, possibly to repair the fabric of the excavation, provide cranes to lift plant and so forth. The main works contractor has no contractual obligation to provide these services to the box-push contractor. The box-push contractor does not have any such resources.

A similar analysis pertains for remediation, though in this case to repair or replace hydraulic rams damaged by flood water will require the expertise of the box-push contractor. An interface violation risk exists if there is no contractual means to retain their services in such circumstances.

However, there is one important difference with remediation – time lag. Prevention is generally a matter of ordinary business for project teams. Containment activities, though generally matters of extraordinary business, are neither entirely unexpected nor unfamiliar. Containment actions are most often tackled with the resources at hand by diverting them from their current use and paying for the work done through variation orders and claims settlements. The general ease with which prevention and containment can be effected does not always extend to remediation. It is a task that emerges after the expertise and resources needed have probably left the project for other work and so cannot be easily brought back. In the example, remediation would require the expertise of the box-push contractors, who may elsewhere be setting up other box-pushes and who have no contractual obligation to repair and reset up their equipment (as happened in the real case).

There is no easy solution to this. It would be a waste of money and grossly inefficient to have the expertise and resources needed to restore assets and damaged, partly completed works just in case they were needed. It would be easier to effect remediation if the expertise and resources needed were part of the same company as the client team, but as luck will have it they will probably belong to other organisations and be committed elsewhere at the time remediation is necessary.

The best an analyst can do is identify the parties that would be needed and point out where the gaps in the contractual commitment may be. The party with the obligation, usually the client, must then accept it is exposed to the costs of obtaining the services of these experts resources at short notice and the consequent premiums it may have to pay.

Funding Block Risk

The objective of the analyst is to get each risk cost allocated to a party under the terms and conditions of their contract, and subsequently to find out the name of the individual within the contracted organisation who has the task of managing the risk so that the project manager can check progress.

Allocating ownership means the names of the organisations that will carry out the work of prevention, containment and remediation are identified in the contracts for their services. The services pertaining to prevention are almost always a matter of ordinary business for project teams. One sees many lists of risk prevention actions that are entirely ordinary business in nature: 'early consultation with…' 'joint working on….' 'approvals sought from…' and so on. These are clearly intended to prevent the unwanted outcome happening by using the resources to hand. The costs of these activities are often covered within project budgets.

However, containment can be more of a problem. The expertise and resources needed tend to be close at hand because the characteristics of the unwanted outcome (what it is and what has to be done about it) are usually similar to those of the ordinary work of the project. Thus, a failed interface in a systems integration project phase tends to require the same engineers and development tools to fix it, as it did to make it. The means for containment are therefore often at hand.

The problem is paying for it. Though the nature of the work required may be ordinary, that it has suddenly emerged is likely to make it a matter of extraordinary business. It is something that is in addition to the planned work schedules. The analyst must therefore make sure that a manager remitted to manage a risk has access to funds to pay for its containment. These might be held by the manager in person, or in a central fund to which an urgent application can be made. But if there is no means to pay for containment, then even if the parties with the expertise and the resources are willing to work together on it, no payment is going to mean no activity. I call these situations funding blocks. Where they exist, they are risks in their own right that should appear in the risk register and model.

A funding block risk is one for which no mechanism exists for a project manager to instruct and pay contractors in a timely manner – to contain or to remedy the unwanted outcomes of risks. Here is an example:

> *The cutting has a history of landslips. Excavating to make it deeper to allow taller vehicles to pass under the over-bridge may precipitate further landslips. The excavation contractor will have the expertise and resources on site to contain and remedy any that do occur, but this would work outside of the scope of its contract. The estimators say the typical total costs of fixing a landslip in this cutting would be between £50,000 and £250,000, but the project manager who could instruct the works has no discretion to issue purchase orders over £10,000. The projects director holds a contingency of £100,000 for the portfolio of projects, of which this is one. Therefore if a landslip with an associated cost of more than £100,000 happens, there is no funding stream to remedy it.*

There will often be contracted liabilities to contain and remedy risk, but the analyst needs to be cautious. Even though the terms of a contract may make a contractor liable to act, the scope of this may be limited to activities that are not sufficiently comprehensive. For example, making a contractor liable to provide all assistance necessary to deal with a flooded excavation does not unequivocally include an obligation to re-hire the pumping subcontractor to do it. The contractor may argue that all it has to do is to set up floodlights, and cordon off the flooded excavation of the site on safety grounds. This is a risk to the funder and so it needs to be incorporated into the analysis.

Remediation is likely to be a case of extraordinary business, though this time with the expertise and resources less likely to be close to hand. As already mentioned, I would not expect a manager to be given funds to retain these resources on the offchance they may be needed. The analyst should notify the client (which presumably employs the project manager) what expertise and resources may be needed, confirming that they are not funded (if that is the case) or explaining how and to what extent if they are. At least then the client is aware of their exposure and the project manager has been afforded a degree of defence against blame for the situation. This may help them tackle the remediation work with less duress.

Occasionally, the analyst may find there is no mechanism in place at all to provide additional funds to managers, should any extraordinary work be required. This may be an oversight, a deliberate strategy, or a ploy to force the affected manager to 'find the money elsewhere' rather than from the client. The analyst will then be in the difficult position of recognising that despite evidence that the client has signed up to a process of risk analysis and management (otherwise why would the analyst have been engaged?), that the process is possibly undermined because it does not provide funding to the people managing the risks at the time they need it. There will be good and bad managers of risks, but despite their best efforts, even the good managers will sometimes be faced with an unwanted outcome – a systems interface that does not work, or an excavation that is full of water. It may not be their fault, but it is almost certainly their job to contain and remedy the situation. To fail to provide them with the means to manage such situations is to threaten the success of the project altogether. It is at this point that analysts must steel themselves and recognise that their duty is not to an individual manager but to the investment itself: to those who are putting up the money, the funders. All projects have risks, and if the commercial strategy of the project executive is not to carry financial provision for them – a contingency fund – or not to allow one that has been created to be spent when necessary, then the funders need to be told. Unwanted outcomes will occur, and any project that proceeds on the basis that its managers are not going to receive funds to overcome them is risking somebody else's investment. Loyalty to the investment can be a tough road for an individual but in analysis, truth and integrity matter.

Reconciling the Extraordinary Costs of Containment and Remediation

I have described the use of primitives to assist the modelling of risk. Every instance of them will have associated with it either a sum of money or a period of time, or both, unless they are 'Not Modelled's'. Having already calculated the cost function, a subsequent analysis of the ownership of the risks that went into it is quite likely to identify some more risks: interface violations and funding blocks, and more possible costs: the extraordinary costs of containment and remediation.

It is important that these are incorporated into the cost function for it to be correct. Cost functions used to inform investment decisions must be post-ownership analysis, as Figure 5.5 shows.

Funders will need assurance that a cost function includes the possible costs of containment and remediation, though within limits, there are so many possibilities. Think of the flooded excavations: the water table might move upwards so that the flooding is permanent and the site abandoned; or the box-push may go off at an angle.

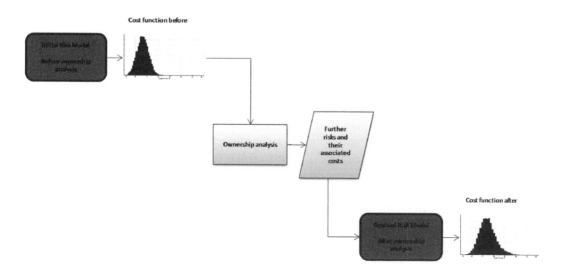

Figure 5.5   Cost function development

The possible scenarios are as many as the imagination will allow, and insurance can be taken into consideration where it can cover the costs. For example

Risk:

There is a risk that the excavation may flood.

Modelling Note:

Prevention: a pumping subcontractor has been retained by the main works contractor and the costs for this are in their price range.

Containment: covered by the services specified in the contract, signed by the pumping subcontractor.

Remediation: covered by the main works contractor's insurance.

It would be sensible to make provision for the project manager's consequential costs (and anybody else's which the funders will have to cover):

Containment: covered by the services specified in the contract, signed by the pumping subcontractor. A provision of two to four weeks per incident has been allowed to cover the time-dependent costs that will be incurred by the project manager.

It would also be proper to declare what is not modelled within the cost function..

Remediation: covered by the main works contractor's insurance. However any consequential costs not covered by this, e.g. starting over but on a different site, are the funders' risk and are not modelled.

and to add in any costs not recoverable from insurance, such as the excess if one is assigned to the policy.

Analysis can grow like Topsy and become the worst enemy to its own objectives. A day's work on this area of risk, the next on another one, and before one knows it a few weeks have gone by. One has created a forest of information within which there are areas they are no longer quite so sure about, especially the workings and limitations of intricate and sophisticated spreadsheet code. This is particularly true for analysis of risk ownership, where the possibilities for ongoing research are legion, viz.: how much will it cost to secure a burnt-out facility? How long will it take to clear the site? How much will it cost to rebuild? How much of this will be covered by whose insurance? When could it re-start? Will a new design be needed?

I think people do this because they are anxious not to be found out to have forgotten something that may in retrospect be important. Also, developing spreadsheets and databases are an absorbing and agreeably confrontation free form of evident industry on one's own part. However, analysts should understand that men are more often brought down by what they themselves have created than they are by their enemies, and so it is with risk analysis.

My advice is not to work and work researching a complete and consistent risk ownership analysis. Instead, understand that clients need action not perfection and therefore an analyst should get the results issued as soon as is practicable. This way, what follows – the management of the risk – can get underway in good time. This will mean that an analyst will probably have to concentrate on the small things. It is the prevention measures that need to be enacted first, and prevention always costs less than containment and remediation. Funders will find it difficult to accept that for the want of a fund that would allow the project manager to bring other pumping contractors onto site, the excavations flooded.

The little things are those that matter: obtaining the small funds that will make a difference. It is better that a dozen risk prevention measures are allocated, funded and enacted than they are not because of continuing research into the containment and remediation of one of them.

Correcting the Estimate and the Plan

There is one further stage. Often, when researching risks and analysing ownership, the analyst will discover costs and tasks that will definitely be incurred and so should feature in the estimate or plan.

In my experience risk analysis is usually started when the development of an estimate and a plan is nearing completion – this is far from ideal. A consequence of this is that awareness of these missing items often emerges too late for them to be incorporated into the estimate and plan, as Figure 5.6 shows.

The analyst must therefore include them in the model, simply for reasons of expediency. Even if the estimators and planners quickly correct their work without the analyst realising it, such that the corrections are then applied twice (once in the spot calculations and once in the spread ones), that is a better outcome than them appearing in neither for two reasons:

1. It errs on the side of prudence – better to be slightly over-funded than slightly under-funded. It's always easier and less public to take back excess funding than to have to find more.
2. Double-counting is much easier to spot in reviews and checks than is the alternative: something that is missing. Errors of commissioning are easier to identify than errors of omission. One knows when one's own car is missing from the car park, but not somebody else's.

This discovery of missing items is natural and not a consequence of superior diligence on the part of the analyst, relative to that of the estimator and planner. It happens because estimators and planners can derive the results of their work from the study of scheme drawings and site plans, whereas analysts cannot look at the same and say what the risks are. The analyst has to get out there and talk to people in order to find out what these may be, and in doing so, will come across things that belong in the estimate and plan. It is a matter of prudence that these should appear in the risk analysis until they do so, and a matter of professional practice that they should be reported to the estimators and planners.

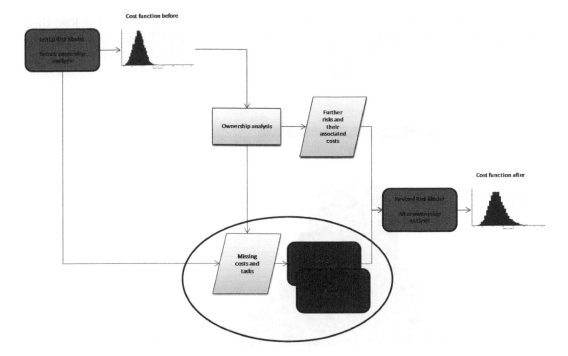

Figure 5.6   Correcting the cost and plan

Using an ICTRA to Inform Risk Sharing

So far, we have derived a cost function from an analysis of the risks associated with the works of a project, and with the transactions for its funding and contracted supplies. A subsequent analysis of the ownership of those risks may have led us to revise the cost function because further risks associated with interface violations, funding blocks, possible costs of containment and remediation, and for items missing from the estimates and plans have been identified. In a perfect world we would now be able to compute a valid cost function using an ICTRA risk model (Figure 5.7).

I have taken the total cost output from the example model, cell B22 of the summary sheet and, in cell B24, added a couple of adjustments intended to represent a typical ownership adjustment.

The spot cost of £15.87M now has (100%–0.4%) probability of being exceeded, and the project has a 95 per cent probability of costing £18.78M or less, a £2.91M increase to cover its inherent risk exposure.

If I can make a reasonable assumption that although an ownership analysis may have been done (and an insight into who should own which risk given), the contracts to ensure this is the case have not yet been signed, then Figure 5.7 is an all-risks picture of the project risk exposure. What every stakeholder will then want to know is how much of it will be theirs, what the funders will want to know is how much of it will they will still have once everything has been said and signed – this is their residual risk.

An ICTRA-derived cost function can give that insight, but before I explain how, it is sometimes useful for an analyst to discuss the utility of the curve with the funders.

Referring to the following figure, if the funders want the project costs to be low, they can enter negotiations with their suppliers (presumably a project management company) about cost values that are towards the left of the X-axis. Figure 5.8 is the same as the previous one but bisected by a line at a cost of £16M.

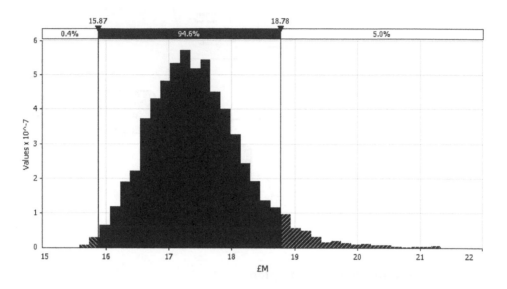

Figure 5.7   The cost function for the example project in density form (as for Figure 4.22, but now assumed to reflect the outcomes of the ownership analysis)

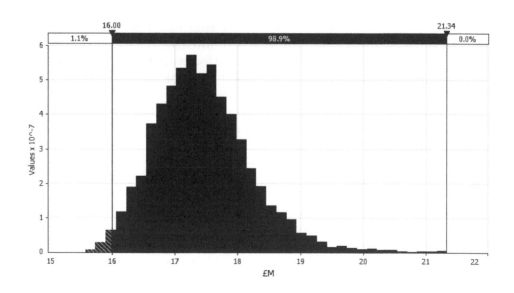

Figure 5.8   A £15.5M price from the lower end of the cost function leaves the funders with almost all of the risk

The scale above the curve shows that there is a 1.1 per cent probability that this or a value less than it will be achieved, effectively no chance at all. The project managers may accept such an offer, provided the funders accept that all of the risk exposure will be borne by them, and subsequently that if any one of the risks that went into the cost function actually happened, the project team could justifiably expect to be given additional funding. The principal drawback from the funders' perspective is that though they may have obtained a £15.5M price for the project, they will not really know what the final cost will turn out to be. Whatever it is though, it will be an overspend.

A further disadvantage to the funders is that it will have to retain capability of sufficient size and expertise to process and assess the claims for risk-related additional funding. It must be anticipated that this will stream up from the project teams, whose job is after all not solely to deliver the asset, but also to make money from doing so. In short, the funders' costs of administering the project will increase, possibly to an extent that the cost estimate part of the cost function will need to be increased.

Against these, the advantage to the funders of negotiating a minimum price is that the unwanted outcomes of the risks may not actually happen, thus the funders can keep the money reserved for dealing with them, provided of course that they had it in the first place.

Alternatively, the funders may not wish to engage in sustained, frequently disputatious talks on claims for risk-related costs. They may instead prefer to leave the project managers to get on with it. In such a scenario the funders implicitly require the project managers to deal with any emerging risks, i.e. to bear them. The funders will be seeking a 'no risk to them price', and so the negotiations would be based on the upper end of the cost function. Here is the same curve bisected at £19M (Figure 5.9). Here, the bar above the curve shows there is a 100%–3.1% = 96.9 per cent chance the project costs will be less than or equal to this.

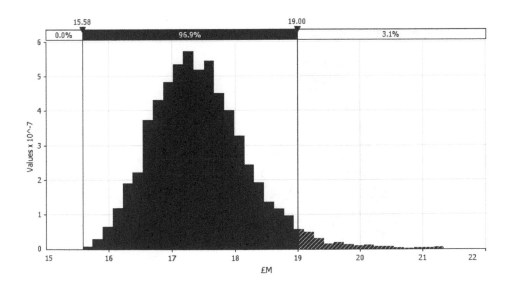

Figure 5.9   A £19M price from the upper end of the cost function leaves the funders with almost none of the risk

If the funders accept this price they need to be aware that their reputation could potentially be damaged. If none of the unwanted outcomes of the risks occur, the project manager will not have to spend any money on containment and remediation (they may even take a chance on not spending it on extraordinary prevention either). They will make windfall profits. These will not be directly attributable to the work it has done but indirectly to the risk it took. Such windfalls may be large enough for them to be politically and socially unacceptable but here is the key point – the money will now be theirs.

There is a further issue funders should be concerned about if negotiations on price are biased towards the delivery team taking risk. The cost function may be packed with risks that are not likely to occur or that can be avoided altogether. If this ploy can be smuggled past the scrutiny of the funders then the advantages to the project managers are obvious. They can be funded for risks that they are unlikely to spend money on that may be usefully reallocated elsewhere to cover unfunded things, such as budget shortages on other projects.

I call this behaviour 'boosting', and it happens when analysts are not acting independently of the parties to the assessment, but instead in favour of a particular faction for whom they are seeking commercial advantage. Or it can happen when the analysts are deficient in intellectual rigour and so identify many risks whose assessments overlap and duplicate each other, or that are modelled in such a way that high-cost potential outcomes are overemphasised (done, perhaps, like shroud-waving in the medical world, to raise the awareness of a problem and of the work done to research it).

Of course what is usually proposed is something in between – a price from the middle of the cost function is negotiated, which suggests some risks will be owned by the project team and some by the funder (Figure 5.10).

This scenario is attractive but it soon collapses when scrutinised. The impression given is that at £17.5M, say, the project managers will take some risk but enough to

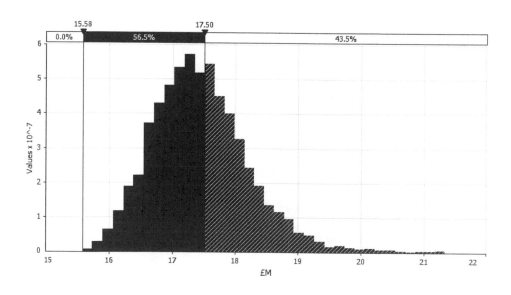

Figure 5.10 An agreed price in the middle of the cost function will leave each party with some risk

leave them with a ~60 per cent chance of delivering the project on budget (or less). The corollary is that the residual 40 per cent of the risk will be taken by the funders.

But say there was one big risk in the model that had a uniform risk–opportunity spread of £0 to £4M. If I plotted this one risk in isolation on the curve it would look like Figure 5.11 – the £0M would notionally contribute to the low end of the cost function, and the £4M to the high end. I have exaggerated what would probably be its Y axis probability of occurrence for artistic effect.

The question is; whose risk is this when it contributes to the exposures of both the funders and the project managers? Moreover, say I had four £0 to £1M risks and the unwanted outcomes on all of them occurred. None of them individually would seem to contribute anything to the top end of the risk exposure curve, but when they all happen, collectively they do, so every risk has the potential to have an effect anywhere on the curve. It is all a matter of which other unwanted outcomes are happening at the same time. Indeed there is a small deception in Figure 5.11 because if the £0–£4M unwanted outcome happened at £4M at the same time as the unwanted outcomes of any other risks, the total exposure would be more than the ~£19.5M.

I therefore come to the conclusion that the individual risk exposures of the parties to a project deal cannot be calculated by divvying up the outputs of the risk analysis, unless it is the special case explained in the next paragraph. Instead, individual risk exposures must be calculated by divvying up the inputs to the analysis and modelling each party's own cost function.

Bisecting cost functions to split the risk between two parties is valid only if all of the risks are to shared, and one party takes liability for (i.e. it funds) all of the exposure of a lower tier (say up to £17.5 M in Figure 5.10) and the other, the top tier (from £17.5M up to approximately £21M in Figure 5.10) no matter which unwanted outcome of which risk it was that caused the threshold of the top tier to be broached.

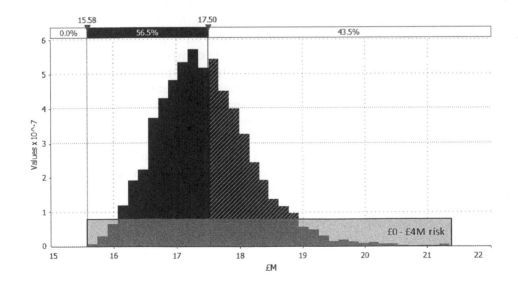

Figure 5.11 A single large risk superimposed on the curve

Thus, if a contractor failed to prevent an equipment failure, taking the costs of containment and remediation over the anticipated £17M threshold, the funder would have to pay (more parties would mean more tiers). This may all be neat and tidy to the analyst, but I doubt it could be sold to the funders. Who then manages which risk? What incentive is there for a project manager whose anticipated final cost has risen up to £17M, all because of failures in risk prevention to do anything more to stop it rising further? It will not be to their cost if it does so the experts and resources allocated to prevention by the project manager may as well be stood down. The project manager may even be anticipating a more empathetic attitude emerging from the other risk-sharing party as it too begins to feel financial pain on what has, because of presumably many failed attempts at risk prevention, been a difficult project so far.

The risk exposures of the parties to a project have to be analysed by partitioning the risks in the model – the inputs – between those that will owned by the funder and those that will be owned by the delivery team. In practice, when negotiating this split there will be some risks that are to be shared, but as I will show this will not create the same problems that sharing the outputs of the risk modelling does. It can be a rational strategy for risks that are beyond the control, i.e. preventive action of either party, be that unavoidably or intentionally (for example – uncertainties about the costs of materials scheduled to be bought some years hence, which neither party may decide to hedge even if they could).

I call what follows the 'funded approach to risk exposure sharing', because my goal is to convince each party that the other has sufficient financial strength to bear its own risk exposure, and as a by-product, to produce a definitive list for each of the risks it has to manage. My intention is that the two parties should come to understand and value what each is doing for the benefit of the other and for themselves jointly.

The funded approach is a direct follow on to the ownership analysis of the previous section. It uses the ownership matrix as the basis for splitting the risks into the two groups, subsequent debate on which may create the need for the third group, the shared risks. The funded approach to risk allocation is thus slightly different to most risk management processes I have seen. These do not allow risk sharing, having a belief that each risk must be owned by one designee. I have sometimes wondered why it is that someone is never the risk manager. It is always someone else, and this, in turn, has led to the conclusion that it is because the 'somebody else' can then be chased for news of progress and that this must pass for work. The remit of risk managers it seems to me is not to manage the risk, but to manage the process of managing the risk.

The risk model has to be split into two: a subset of risks that are to be owned by the funder; and a subset by the delivery team (I will work with just these two parties, though the approach is scalable). This is easy enough to do by adding a tag in an adjacent cell in the spreadsheet containing the risk model, setting this to 1 for a funder's risk or 2 for a delivery team risk, and using Excel SUMIF formulae to generate the cost function for each party (see column H and cells H42 and H46 in the risk model extract in Table 5.5.

Table 5.5   An example of risk tagging

Row/column	A	B	C	G	H
1	Risk no.	Synopsis	*Modelling note*	Sample	Funded allocation tag: 1 Client 2 Project manager
2					
3	1	The p way relaying estimate is a best guess. Between –1 and +1 mile more could be required @ £1M per mile, but with reducing probability.	*Spread Q (continuous): quantity modelled as triangular –1.0.1*	£166,456	2
4					
5	2	Some existing trackside equipment cabinets may have to be moved to another site. Funding for 6 has been included in the estimate, but up to 4 more may also have to be moved, depending on the final layout of the track.	*Spread Q (discrete):*		
6		The cost of relocating an REB depends on the number in the subcontract, from £40k per unit down to £25k per unit.	*Spread R (covarying):*	£132,557	2
7					
8	3	The 12 under track orange tube cable ducts might instead have to be 2 UTXs at a cost of £150k more.	*Unresolved option (permutation of 1 from 2), @ 50/50 for the cost difference.*	£0	2
9					

Table 5.5   Continued

Row/column	A	B	C	G	H
10	4	There are 20 overnight closures planned for rural crossings on the first weekend in November for the commissioning of automatic barriers. Historically one in ten of this type of possession fails to complete, in which circumstances we pay a £2500 fixed penalty to the local authority.	*Random failures*	£10,000	2
11					
12	5	There is a probability that any of up to six pieces of land will need to be bought. Each piece of land has a different value. The Property Department say purchase will only be needed if a specific request to divert the railway away from a hospital is agreed. It is felt that there is 75% chance that the request will not be granted.	*Unresolved option (permutation of N from 6) @ 25%*		
37				£0	1
38					

Table 5.5   *Concluded*

Row/column	A	B	C	G	H
39	9	The works of the project must not alter the risk to the public. Any works that do must be compensated by additional works that restore the status quo.	C2L	£50,000	1
42					Total client risk = SUMIF($H$3:$H$39,1,$G$3:$G$39)
46					Total project manager risk = SUMIF($H$3:$H$39,2,$G$3:$G$39)

By tagging a risk to a party I am making an assumption that that party is or will become liable for the costs of prevention, containment and remediation associated with that risk. This is based on the results of the analysis of ownership matrix, though I accept this may be altered by the contractual negotiations in due course.

Where the ownership matrix has interface violations and funding blocks I recommend these are recorded in the risk register, modelled just like any other risk and given to the ownership by the funders. Being at the top of the procurement chain and so being the paymasters, the funders have the means to insist these situations are remedied, and ultimately to refuse to finance the project if they are not.

Some risks will be neither 1 nor 2, i.e. the risks that are to be shared. They will need more sophisticated modelling that reflects how the sharers intend it to be done. There are two basic ways to do this: the vertical slice and the horizontal slice. When running, a risk model generates possible sample values for a modelled risk according to how it has been encoded, each of which is a possible financial consequence of the risk. In a vertically sliced risk share, an agreed proportion of each sample is allocated to each party – 50/50 or 60/40 for example. Here is an example of a vertically sliced risk, illustrated in Figure 5.12, and its encoding (Table 5.6).

There is a risk of emerging legislation imposing costs on the project providing enhanced protection requirements that are intended to prevent damage to third-party assets by the project.

There are three possible degrees of protection, so the risk is modelled as an emerging cost chain. The estimators advise the costs could be between £100–£200k for protection to Level 1; £150–

£300k for Level 2; and £500–600k for Level 3. The probabilities of each Level are respectively considered to be 25 (for no additional protection required), 35, 35 and 5 per cent.

The uncertain nature of the legal requirements has led the funders and the project managers to agree to share this risk on a 75 per cent funder–25 per cent project manager split of the emerging costs: the funder will cover the additional materials and labour, while the project managers will manage the work.

The project managers have accepted an argument that they were aware of the emerging legislation and should have allowed for the additional protection in their pricing. They have consequently accepted a 25 per cent share of the exposure, effectively managing the additional work at their own cost.

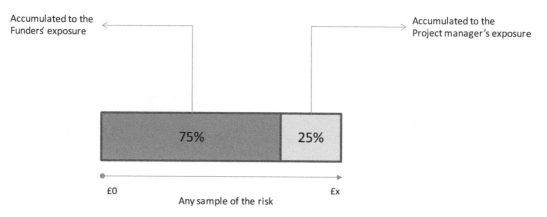

Figure 5.12 A vertical slice risk share

Table 5.6   A vertical slice

Row/column	A	B	C	D	E	F	G
1	Synopsis	Modelling note	#1: Cases	#2: Which case?	Sample	Slice 1: funders	Slice 2: project managers
2	As set out in the preceding explanation.	As set out in the preceding explanation.				0.75	0.25
3			£0	= RMT/ discrete ({1,2,3,4}, {25,35,35,5})	= CHOOSE (D3,C3,C4,C5,C6)	= E3 * F2	= E3 * G2
4			= RMT/uniform (100000,200000)				
5			= RMT/uniform (150000,300000)				
6			= RMT/uniform (500000,600000)				

Should the unwanted outcome occur in a vertical slice, both parties will pay for its remedy according to the percentage split they have agreed (or are in the process of negotiating – they will probably want to see their individual total exposures before agreeing).

In a horizontal slice (Figure 5.13 and Table 5.7), one party bears all of the financial consequence of the risk up to an agreed value. Any cost above this will be borne by the other party. I will use the same risk again, but this time imagine that the funders have insisted the project managers ought to have tendered for the project cogniscent of any emerging legislation. As a result they have refused to underwrite anything below what they see as a fair level of exposure to this risk.

There is a risk of emerging legislation imposing costs on the project. It is intended to prevent damage to third-party assets by the project.

There are three possible degrees of protection and so the risk is modelled as an emerging cost chain. The estimators advise the costs could be between £100–200k for protection to Level 1; £150–300k for Level 2 and £500–600k for Level 3. The probabilities of each Level are respectively considered to be 25 (for no additional protection required), 35, 35 and 5 per cent.,

The funders have decided that the project managers must bear the first £500k of this risk but that they will accept any increment above this.

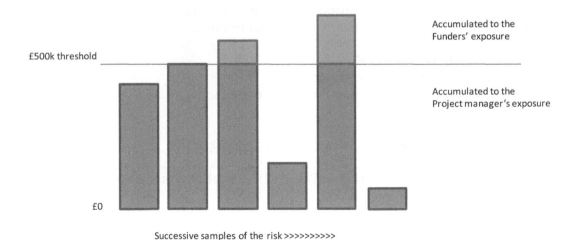

Figure 5.13 A horizontal slice risk share

Table 5.7   A Horizontal Slice

Row/column	A	B	C	D	E	F	G
1	Synopsis	Modelling Note	#1: cases	#2: which case?	Sample	Slice 1: Funders	Slice 2: Project Managers
2	As set out in the preceding explanation.	As set out in the preceding explanation.					
3			0	= RMT/ Discrete ({1,2,3,4}, {25,35,35,5})	= CHOOSE (D3,C3,C4,C5,C6)	= IF (E3>5000000, 500000,E3)	= IF (E3>500000, E3 - 500000, 0)
4				= RMT/Uniform (100000,200000)			
5				= RMT/Uniform (150000,300000)			
6				= RMT/Uniform (500000,600000)			

The Excel code for the two slices is easy to write, as is the tagging that discriminates between risks owned by the project manager and risks owned by the funder. Both can be added in an overlay to an existing model without the need to fundamentally alter it.

One further type of risk share may be appropriate, and indeed its introduction may unlock a difficult risk sharing negotiation. This is the functional split in which each party agrees to pick up some risk costs but not others. For example:

- Labour costs but not materials.
- Further design but not further construction.
- Re-testing and re-commissioning, but not preceding software modification.
- All costs except third party.

It is easy enough to devise Excel code to model such a risk share:

Risk cost = RMT/Discrete ({1,0}, {25%, 75%}) * RMT/Uniform (£50000, £75000)

Party 1 cost = 60% * Risk cost

Party 2 cost = 1 – Party 1 cost

and such like, with such increased sophistication as may be necessary.

We are now (apparently) in a position to run the model and calculate the discrete exposure curves functions of each party (or, if these include their respective cost elements, the discrete cost functions). However, there is one final shared risk, though I hesitate to call it that because it is more of a by-default scenario. Say we now have a pair of cost functions, one for each party as shown in Figure 5.14.

Here the funders have accepted a cost function of their own, approximately £6M–7.5M. The project manager's cost function is approximately £9M–11.5M. But the

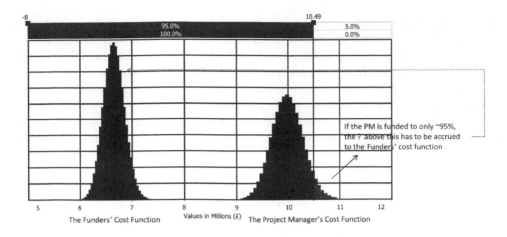

Figure 5.14 Two cost functions: the funders' and the project manager's

money project managers spend is always somebody else's and so the funders will pay for this function too.

Analysts need to be alert to the possibility that the funders may strike a deal with the project managers to fund them up to, say, a 95 per cent level on the project manager's curve. The question is then what happens to the residual 5 per cent? It may be that the project manager is prepared to take the quite reasonable chance it can take on all of the risk, i.e. 100 per cent of its exposure, for 95 per cent of the funding. It would show it has an appetite for risk and the funders may well be pleased with such a deal.

However, the detail of the deal may be different. It may be that the project manager will be funded up to 95 per cent, but any incremental expenditure above this will have to be met by the funder. If the terms of the deal are like this then Figure 5.14 is wrong, as it shows any excess on the project manager's function is not added back into the funder's. It may be even slightly more complicated in that above this limit of agreed funding, the increments are shared between the funders and the project manager in a pain share. Perhaps the manager will only receive 25 per cent of the overspend, in which case the additional cost accrued to the funders' cost function will have to be factored accordingly.

Furthermore, a pain/gain share may be contracted in which the amount of any under-spend below the limit of agreed funding is split between the funders and project manager (who will get its share as a bonus). Again, the funders' cost function will need to be revised to reflect this possible opportunity for money to be returned to them.

In conclusion, once the two parties have decided how the cost function will be split, i.e. what the deal for the project will be, the analyst should adjust the model to reflect their intentional agreement. They should also re-run it because even though the project manager's part will probably be transformed into a spot cost, there may be mechanisms in the detail of that transformation that will alter the funders' cost function. Once all has been done, the correct cost curves for each party can finally be computed.

The modelled deal now has to be reflected in the terms and conditions of the legal framework. Though it is not the role of the analyst to draft the legal documents, it is their responsibility to advise the appropriate legal and commercial departments of their analysis. This is so the risk allocation, modelled and presumably agreed by parties, can be included in the commercial framework. It would be naive of me not to point out that more times than not, legal teams will proceed with the drafting of commercial frameworks without heed of the modelling results. If they did, then there would probably be far fewer transactional risks than there are. Later revisions to risk models are the norm.

There is one more insight the cost functions give that funders should consider before signing a contract with the delivery team. This is relevant in situations where the project manager's function has been transformed into a spot cost – a fixed-price deal – with no recourse to the funders if an overspend arises sometime in the future. Having some familiarity by now with the overall cost function, they will know how much the project could cost the project manager if it all goes wrong. This is the area upside of the price on the project manager's cost curve.

In Figure 5.15, the project manager has agreed to deliver their part of the project for a fixed price of £10.5M. If the terms of the deal are such that if its costs go above £10.5M, to slightly over £11.0M, it has no recourse to the funders. It will be prudent for the funders to check that the contractor has a sufficiently strong balance sheet to absorb what could be a loss if the unwanted outcomes of the risk occur. If the contractor does not, then perhaps the award of the contract should be reconsidered. If it does, then perhaps the funder should ask that an element of the contractor's funds of sufficient size, approximately £0.5M in this case, be held as a bond, or by way of a bank guarantee.

Finally, it is important that any risks not owned by anyone are modelled in a separate cost function, created specifically for unallocated risks reported separately to the client, whose responsibility it then is to decide what to do about them.

It is only after all the work implicit in Part I has been done that I suggest a project has been properly funded; the risk identification, the modelling of the cost function, the splitting of the cost function, the incorporation of this in the contracts. Because of this, I think its chance of being delivered without overspending or overrunning is better than it may have been.

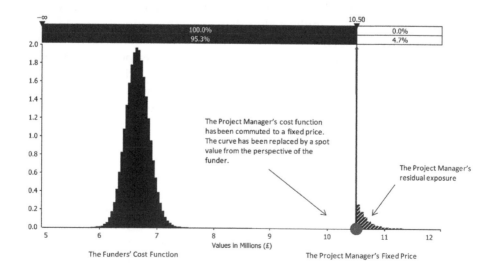

Figure 5.15 The funders' cost curve and the project manager's fixed price and residual risk exposure

Let me bring this section to a close with a summary.

1. A cost function can be modelled for a project using an integrated cost and time risk assessment that uses a rich set of modelling primitives.
2. Who is best placed to own a risk can be determined by the analysis of which party has the expertise, which party has the resources and which party has the obligation to prevent, contain and remedy each risk.
3. This analysis may find further risks: the transactional, interface violation and funding block risks that need to be included in the risk model.
4. The cost function can then be shared between the parties to a project deal. This is based on the ownership analysis but may necessitate further revisions to the risk model to accommodate discrete risk sharing, leading to yet more risks arising from pain and gain arrangements and limitations on expenditure.
5. The final cost functions for the parties can then be computed.

The funders will now have an increased confidence that the project has the right amount of funding to be delivered on budget:

*   They will know when the project is likely to be completed.
*   They will know what their residual risk exposure is, and what risk the project managers are taking.
*   They will know that the project manager has sufficient financial strength to deliver the project on time and on budget.

This perfection of knowledge is unlikely to survive the vicissitudes of project life as the scheme goes forward, so there is a continuing role for the analyst of monitoring, reviewing and revising their work on behalf of the funder. The position will change and a need to refinance a project may be foreseen or as is more often the case, the period of exposure to a risk has passed without incident so some of the risk funding can be returned for investment in other projects. Unforeseen risks will emerge, the impact of which will need to be understood quickly. Some parties may leave the funders and yet others may join, the scope of the project adapting to emerging needs and so on, but essentially that start with a blank sheet of paper is, for the analyst, over and done. What follows in terms of remit from the funders will be largely maintenance and modification.

However, a new role requiring a blank sheet of paper and fresh thinking does arise for the analyst provided their focus changes from informing the funders to supporting the project manager.

From the perspective of the funders, the project manager's cost function will have been transformed to a price, but to the project manager that spot price remains a curve. The challenge now for the analyst is how they can help the manager deliver the project on time and to budget. Is it now just a matter of grip, drive and kicking butt, or can risk analysis play a part to good effect as well?

6 Using Risk Analysis to Derive the Risk Management Strategy

Introduction

There is an old saw that says a consultant is someone who tells you the time with your own wristwatch, and this is very easy for risk analysts to do. Others explain what risks they are facing, and after some paperwork, the analysts present them with a report on the risks. A few weeks later, this is followed by a finger-jabbing inquiry into progress on risk reduction action. What the added value of such a limited process is to the people managing the risk eludes me. I suspect the practice survives largely because the people explaining what the risks are generally do not take notes of their discussions with the risk people.

Therein is the weakness of the risk management process – it is not sufficiently critical to the people managing the project because its outputs will neither alter nor influence what they have to do. Project managers may assert this is not the case, but my experience is that risk analysis work is often used to *confirm* intentions rather than to alter them. No matter (for the risk analyst that is), the process of risk management has a secure place in the pantheon of good project management practice. It is clearly a good thing and once included in the practice of a business it will set about stimulating the writing of manuals, the ordering of IT applications, the recruiting of hierarchies, the development of training and the holding of meetings. But when it comes down to the value its intended beneficiaries put on it, hand on heart, if I stopped or never even started up a risk management process, I cannot say the people managing the project would desperately attempt to make it not so. This suspected disinterest would not pertain for the companion processes of cost control for the estimate and progress monitoring the plan. Stop or not even start these and people would be incredulous, asking how they are supposed to manage the project without them?

Just to clarify, I am writing here about the team managing the project, not the people funding the project. The latter are always desperately keen to know the results of the risk work. It is when the target audience for the risk work swings away from the project funders and over to the project managers that the interest for risk analysis starts to wane (until such time as the project runs out of money or time, of course).

This ought not to be the case, because there is much that analysis of the risk can do to make the management of the risk more effective, and in the following sections I hope to explain how.

To see how, consider the following figure.

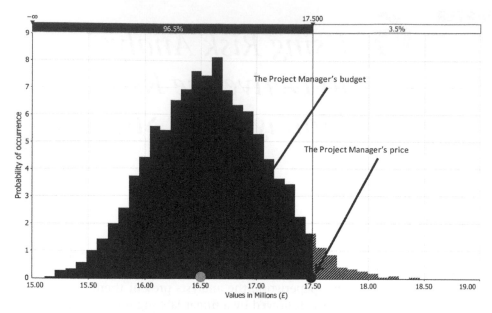

Figure 6.1 A cost curve fixed price and budget markers

Figure 6.1 shows a project cost curve. The black dot to the right marks the value of a fixed price deal to deliver it, £17.5M. However, in practice this is never the amount given as a budget, because various contingency funds will have been appropriated from it for retention by the programme director, the CEO and the main board to name but a few. The green dot to the left is what is handed over to the PM, £16.5M.

In this example, the figure in the budget is £1M below the fixed price of £17.5M, which at less than 6 per cent of the fixed price is not an unreasonable target for savings to impose on a delivery team. What Figure 6.1 shows is that there is slightly less than a 50 per cent chance that the project can be delivered for an outturn cost less than or equal to this budget. This is the proportion of the area under the curve to the left of the green dot. What *was* a reasonable target, a 6 per cent saving, may now look decidedly less so to the delivery team project manager.

The message of the figure, that there is approximately a 50 per cent chance the project will overspend, may sometimes be presented by analysts as a challenge to the project manager. This is not a good approach to presenting bad news. Project managers deal every day with people who are a lot more hard-boiled than analysts and may tend to view any waspish doom-saying by an analyst as coming from, well, a wasp that is going to be dealt with in one of the time-honoured ways. Project managers will be working to the budget they have been told, and having an analyst pop up to say there is a significant probability (the speckled area to the right of the budget) that they will not be able to do deliver the project for this sum will, quite understandably, be regarded by them as an experience to be avoided.

However. a risk analyst can offer two things that should to have the project manager asking to be told more. The analysis behind the Figure 6.1 shows the size of the contingency fund provided to underwrite fully the risks the project manager is trying to control. That the project manager's employer, the project management company, has agreed to a price less than the value of the full exposure for taking on the risk is not the project manager's responsibility. Thus, first, the analyst can use the figure to inform the project manager of the size of the contingency fund.

Though a situation in which a draw-down from a contingency fund is required could be considered career-limiting by the project manager, this need not be the case if the risk analyst is able to remind the project management company that it, not the project manager, took on the risk. There are often situations in project management where good men and women are deemed to have failed because the projects they are managing start to overspend and run late. This is nonsense. A project manager does not choose the risks their employer contracts them to bear, and not all risk prevention actions are successful. It may well be the case that the person most able to rectify a project starting to run late and go over budget is the current project manager. Therefore an analyst who understands the risk, and who knows it is being managed well, can defend the project manager where this would be in the best interests of the funder.

Secondly, though the situation represented in Figure 6.1 does show there is a high probability of an overspend, a significant area under the curve is not speckled – to the left of the budget – and an outturn cost in this area would be deemed a success by everybody involved. The risk model that generated the cost function is a set of equations and as such, it should be capable of being solved to find a set of circumstances that need to prevail for the project to be delivered on budget or for less. If these circumstances can be made to happen then the risk analysis will have produced a solution to managing the risk, i.e. it will have provided the manager with an analysis-based plan for managing the risks.

I used the phrase 'plan for managing the risks' to avoid confusion for the moment with the commonly used term 'risk management plan'. This is often a document that contains a series of statements about processes, empowerments, timetables, responsibilities and deliverables, but no solution for managing the risk to achieve measurable goals. The plan I have in mind does. It comprises a set of specific actions derived from the risk model that will, if successful, let the project be delivered on time and to budget. Some will call this a 'risk-reduction action plan' but in what follows I will also now use the title a 'risk management plan'.

I use the terms 'manage the risk', 'control the risk' and 'treat the risk' to mean the same thing: it is the activity to be undertaken if intended either to close a risk, to limit its potential impact (stop it getting worse), to reduce its potential impact (make it get less, either by reducing the probability of it happening or by reducing its possible effect). This activity can be a matter of ordinary business (a prevention activity, e.g. agree a specification for an interface) or one of extraordinary business (a containment activity not in the plan for the project, e.g. make safe a collapsed tunnel).

An analysis-based risk management plan can generate a very different risk-reduction strategy from that derived from the traditional approach of managing the risks as a high to low priority based on the evaluation of their potential impacts. An analysis-based plan may ignore a high risk on the grounds that no amount of management will affect the probability of its occurrence or impact. It may instead propose resources be directed towards a group of smaller risks, none of which appeared individually in any top 10 ranking, but that can be reduced or eliminated by one or more practical actions, the potential benefits of which are supported by evidence from the risk model.

The first step that needs to be taken when deriving an analysis-based risk management plan is to produce the cost curve that is specific to the risks the project manager owns. This does not necessarily mean contractually because that is more likely to be their employer, but those for which the manager is the risk-reduction action owner.

As explained earlier, this is easy when using the Excel SUMIF formula if the manager's risks have been tagged with a marker in the risk model. It is better to use the full model with tagging rather than to extract a part-model containing only the manager's risks. This way, the manager can be reassured that risks they may be concerned about have been taken into account and that they are tagged to another owner. I have models that use many tags covering a number of funders, managers and contractors, who share the exposure of a single large project. In such a scenario, being able to show one party that a risk is in another party's cost function can often provide a welcome reassurance that the risk is covered somewhere.

However, having computed a manager's cost curve, an analyst is most likely to be first asked what the top risks are, and it is this I will write about first before moving on to the risk management solution.

Ranking the Risks

The first thing an analyst usually does to help the manager manage their risks is to identify the ones that pose the greatest potential threat. This is due to the reasonable assumption that identifying them will lead to their treatment being given priority and hence that the management of risk will, in some sense, be efficient. However, there are some weaknesses to this that I will explain after discussing the ranking options.

I use five ranking techniques, none of which is any better than the others. The rankings are: mean, correlation, urgency, trend and leverage. Each provides a slightly different insight into the characteristics of the risks that will help the analyst suggest how the manager can control their overall risk exposure efficiently in one way or another.

Ranking 1: Mean

The mean used here is the arithmetic mean: the average value of the data samples taken, computed either in Excel or RMT for each risk as the modelling run completes (Table 6.1).

In Table 6.2 the average of 5000 samples, each of a set of risks, is calculated in the second row. These are plotted in descending order of magnitude using standard Excel to give the Tornado Graph of Figure 6.2.

The most common units for the average are money and time, though this need not always be the case if the risk model has been based on another currency – the number of accidents, the additional quantity of material, the extra man hours of labour or the number of lost fare-paying passengers.

The idea behind average value ranking is that giving priority of reduction action to the largest risks will be an efficient strategy for their management. But the method is flawed, it can under-value some kinds of risk. Look at Risk 1 in the figure, the low valued one towards the bottom of the tornado. The synopsis and modelling for Risk 1 in the model used to generate the samples were (Table 6.3) nd it has a average of £4 (from row 2 of Table 6.2 – which is noise in the calculation because it should theoretically be £0).

This risk is equally balanced between more and less rail track being required. If, instead of modelling it as –1 mile to +1 mile, I had instead decided it was risk of +1 mile more paired with a separate opportunity of –1 mile less, then the risk part would have had an average value (derived the triangular function I used) of £333k. Thus, it would have been the biggest risk and appeared at the top of the tornado, while the opportunity part would have been at the bottom below Risk 8 (a negative average indicates the risk is instead an opportunity).

Management experts will sometimes argue that risks should be kept separate from opportunities because the skills needed to realise an opportunity are not those needed to avoid a risk. If the two are merged so that the risk is offset against the opportunity,

Table 6.1   A set of samples from a risk and the mean

Sample number	Risk samples: project management charges are per accounting period for the duration of the project and so are a time-dependent cost
1	£97,564
2	£174,345
3	£94,015
4	£269,612
5	£177,965
6	£319,440
7	£212,772
8	£387,804
9	£112,537
10	£74,682
↕	↕
9991	£195,622
9992	£67,403
9993	£260,558
9994	£118,925
9995	£139,423
9996	£156,960
9997	£214,444
9998	£212,575
9999	£75,101
10000	£350,889
average	= AVERAGE($B$2:$B$10001)
	£210,698

Table 6.2 A set of risk averages

Average -/- sample	Risk 1	Risk 2	Risk 3	Risk 4	Risk 5	Risk 6	Risk 7	Risk 8	Risk 9	Risk 10	Risk 12
	£4	£59,000	£75,000	£5,000	£52,521	£276,591	£74,986	-£101,552	£50,000	£63,340	£210,789
1	£293,628	£91,091	£0	£0	£16,760	£333,709	£0	-£100,000	£50,000	£59,449	£165,402
2	-£404,370	£65,453	£0	£5,000	£0	£406,744	£0	-£100,000	£50,000	£72,177	£229,537
3	£666,888	£35,930	£150,000	£5,000	£0	£366,168	£0	-£90,000	£50,000	£64,446	£177,606
4	£219,953	£34,623	£150,000	£0	£0	£349,558	£267,691	-£120,000	£50,000	£64,242	£246,863
5	£266,040	£110,178	£150,000	£7,500	£0	£360,066	£0	-£110,000	£50,000	£60,576	£153,817
6	-£488,576	£87,745	£0	£0	£0	£387,270	£0	-£100,000	£50,000	£72,907	£279,722
7	£12,283	£36,420	£0	£10,000	£0	£368,024	£0	-£100,000	£50,000	£60,835	£281,291
8	-£270,101	£104,297	£150,000	£10,000	£0	£0	£0	-£100,000	£50,000	£63,341	£322,233
9	£174,924	£62,148	£0	£7,500	£0	£409,836	£0	-£90,000	£50,000	£61,433	£194,669
10	-£298,474	£85,746	£0	£2,500	£0	£326,529	£0	-£110,000	£50,000	£70,322	£202,051
‡											‡
4991	£221,316	£35,005	£150,000	£2,500	£45,275	£350,494	£0	-£110,000	£50,000	£56,017	£197,861
4992	-£158,463	£0	£0	£7,500	£0	£372,397	£0	-£110,000	£50,000	£70,399	£96,482
4993	£8,258	£103,063	£0	£7,500	£393,522	£422,674	£335,260	-£100,000	£50,000	£66,342	£292,616
4994	£63,107	£63,940	£150,000	£2,500	£0	£377,670	£0	-£100,000	£50,000	£63,379	£163,756
4995	-£751,761	£36,265	£0	£10,000	£0	£0	£0	-£100,000	£50,000	£70,611	£188,746
4996	-£162,677	£103,787	£0	£5,000	£0	£385,136	£0	-£110,000	£50,000	£58,317	£314,763
4997	£355,584	£91,576	£0	£5,000	£0	£356,871	£0	-£90,000	£50,000	£68,390	£184,530
4998	£700,416	£34,428	£150,000	£7,500	£0	£333,262	£0	-£100,000	£50,000	£71,108	£119,255
4999	-£673,880	£64,860	£0	£0	£0	£333,191	£0	-£100,000	£50,000	£57,843	£267,078
5000	-£54,891	£34,676	£150,000	£5,000	£0	£397,945	£0	-£90,000	£50,000	£64,224	£237,356

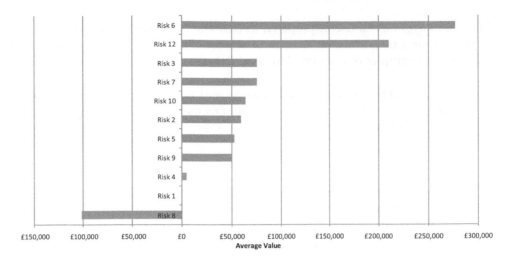

Figure 6.2   A tornado graph from a set of risk averages

Table 6.3   A combined risk and opportunity

Synopsis	Modelling note
The quantity of rail track needed is a best guess. Between −1 and +1 mile more could be required @ £1M per mile, but with reducing probability.	*Spread Q (continuous): quantity modelled as triangular −1.0.1*

no management skill of either sort at all may in consequence be seen to be needed, with potentially serious results.

Generally, the average rankings can be misleading if risks have been offset by opportunities. Though it is perfectly good practice to model them as shown in Table 6.3 because the overall spread of exposure will be unaffected and because it makes for a more concise model to do so, I recommend they are kept separate in the rankings. In the Table 6.3 example it would be better to compute the average of the negative samples (those that reflect the opportunity) and the average of the positive samples (those that reflect the risk) ranking each separately.

Furthermore, because the average is easy to understand, behavioural problems often arise. Some risk reviews I have attended have had forceful characters present who have often decided a plus value to a risk quantification is an affront to their ability to show grip, drive etc. They counter it with a minus value that reflects their grip, drive etc., for example: −10 per cent to +25 per cent, −100 to +1000 and so on, where the −10 per cent and −100 reflect ambition. This is a classic case of taking a heroic view, the consequence of which is that the two opinions, heroic minus and pessimistic plus, cancel each other out to a degree with the possible consequence that major risks may be lower in the ranking than they ought to be. In such circumstances, ranking the average opportunity separate to the average risk will let the analyst show the funder the true picture.

Mean rankings have a useful property, in that the combined weight of two or more risks can be found by adding up their mean values. This is useful where a risk register contains many examples of the same low-level risk, for example:

Each weekend of 20 weekends where outdoor winter work has been planned may be disrupted by snowfall. The contractors may have to be re-hired for more weekends in the spring in order to complete the project.

This could be modelled as 20 discrete risks: Risk 1: weekend 1 production lost; Risk 2: weekend 2 production lost and so on. But because the probability of snow falling on the weekend is low, each risk would be low in the rankings. This could be misleading if over the 20 weekends at least one or two are very likely to be lost to snow with significant cost consequences. The potential problem could have been identified in the risk ranking if the means of the discrete risks had been summed so that the combined risk was given a more prominent rank. Analysts should therefore always check the lower levels of mean value rankings to see if a much larger risk has been disguised by its dispersal amongst several small ones.

Ranking 2: Correlation

Consider the plan in Figure 6.3 and its risked-up emulation.

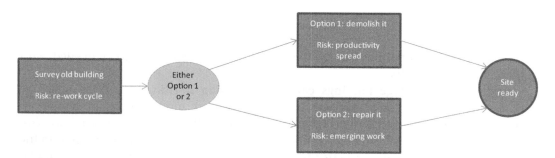

Figure 6.3   A plan for some site preparation works

A site is to be prepared for construction purposes. There is an old building on it that may be repairable and re-used. On the other hand it may have to be demolished, it all depends on the outcome of a survey. Table 6.4 is an emulation of the above plan with some risks incorporated.

Each activity has a spot duration in column H and spread duration in column L. These reflect the risk description in column G. The logic of the plan is encoded in column K and the overall duration derived in cell M6. This is a risk emulation so the spread durations in L are used to calculate the overall duration of the plan. Those in H, are not.

Table 6.4   The plan emulated, with risks incorporated

Row/ column	F	G	H	I	J	K	L	M	N
1	**Task**								
2		**Risk**	**Spot duration**	**#1**	**#2**	**Start**	**Spread duration**	**#3**	**Finish**
3	Survey building	Re-work cycle: survey is inconclusive and further surveys may be required. 50% chance of a further week.	4			1	= 4 + (RMT/ Discrete ({1,0}, {50,50}) * 1)		= K3 + L3
4	Either repair of demolish the building	50/50 mutually exclusive probability		Which task?	= RMT/ Discrete ({1,2}, {50,50})				
5	Repair building	Emerging works. Allow for a further 4 weeks	16			= N3	= RMT/ Uniform (16,20)	= if(J5 = 1, L5,0)	=K5 + M5
6	Demolish building	Productivity spread: allow +–1 week.	4			= N3	= RMT/ Uniform (3,5)	= if(J5 = 2, L6,0)	= K6 + M6
7	Site ready for next stage	Milestone. Zero duration	0			= MAX (N5,N6)	0		= K7 + L7

The outcome of the survey is not known and so the building may be either demolished or repaired (see J4). On the basis that there is no strong opinion among the project team about which it will be, each outcome has been given an equal 50 per cent chance. Logic in M5 and M6 models the mutual exclusivity of demolition or repair. The duration of the task that is not done is set to zero so that it does not influence the overall duration.

Running this model produces means for each risk as illustrated in Table 6.5.

Ranking by the mean is obvious, as is the form the tornado would take:

- repair, survey, demolish,

in descending order, with the dimensionless 'which task?' excluded because it does not have numbers of weeks associated with its value of 1.5 (which is caused by being 1 or 2 on a 50/50 basis). This exclusion could be significant.

The ranking is different if we look at an alternative measure, the correlation between the samples of the risks and the samples of the overall project duration (Table 6.6).

Table 6.5   Average time for each task in weeks

	Survey building	Which task	Repair building	Demolish building	Site ready
Average	5	–	18	4	16

Notes: 'which task' is a logical switch and is duration-less. 'site ready' is a milestone and also duration-less.

Table 6.6   Correlation coefficients

	Survey building	Which task	Repair building	Demolish building	Site ready
Average	5	–	18	4	16
correlation	0.14	0.99	0.05	0.02	1.00

Correlation is a dimensionless measure and so it allows the logical switch, duration-free element 'which task?' to be ranked alongside the correlations of risks that do have duration parameter. By this measure the ranking in descending order is:

• which task?, survey, repair, demolish.

On the basis of this insight, placing a decision point in the plan that introduces the probability of repair can be seen to have increased the risk of the project overall, and so the project manager or funders may decide that demolition is the only viable course of action, thus excluding the choice.

To see how correlation works, take a look at the table of samples in Table 6.7 from a cost risk model. A set of eight concurrent samples of each is shown for each of the three risks taken from a cost risk model, as is the contemporaneous overall total for all of the risks.

Table 6.7   A set of samples from three risks

Sample number	1	2	3	4	5	6	7	8	Mean	Correlation
Samples from risk 1	£50.0k	£45.0k	£38.0k	£51.0k	£53.0k	£40.0k	£46.0k	£47.0k	£48.1k	0.39
Samples from risk 2	£80.0k	£81.0k	£83.0k	£79.0k	£80.0k	£78.0k	£80.0k	£80.5k	£80.0k	−0.29
Samples from risk 3	£10.0k	£11.0k	£8.0k	£8.0k	£8.0k	£9.0k	£11.0k	£7.0k	£9.4k	0.98
Samples of total cost	£1.2M	£1.3M	£.9M	£1.0M	£1.0M	£1.1M	£1.4M	£.9M	£1.1M	1.00

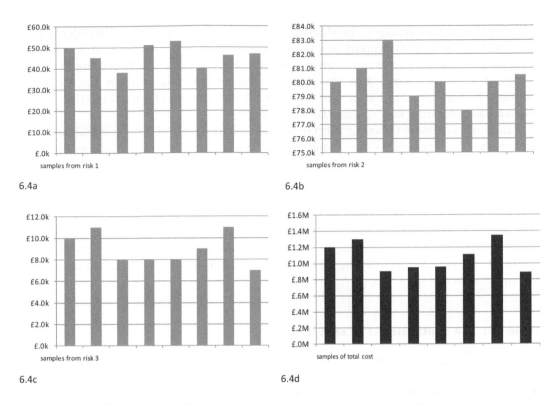

6.4a

6.4b

6.4c

6.4d

Figure 6.4   A set of samples from three risks and the overall total of these and many other risks

If the four sequences are plotted as column charts (Figure 6.4) then although Risk 2 has the highest mean, followed by Risk 1 and Risk 3 (see the penultimate column in Table 6.7), if we look at the change in value from sample to sample, the profile of the sequence of changes in Risk 3 can be seen in the profile of the overall output, i.e. whatever variation takes place in Risk 3 can be seen in the overall result. The overall result is therefore more highly correlated with Risk 3, and much less so, if at all, with Risks 1 and 2. On the basis of correlation – the degree of similarity – Risk 3 must be a bigger risk than 1 or 2.

An analogy may help see the difference between them:

If I were to ask you to hold out your hand and then place a one kilogram block on it, it would probably move downwards. But I could also make it move downwards with something much lighter, a pin. The block has weight (average) but the pin has influence (correlation).

The statistic that measures the degree of similarity or influence, the correlation coefficient, is provided by Excel:

= CORREL(B3:I3,$B$9:$I$9)

The values of a set of correlation coefficients can easily be plotted in Tornado form using Excel's standard bar charts.

Note that, unlike averages, correlation coefficients cannot be summed in order to give an indication of the joint impact of a set of risks. If an analyst suspects that an external event such as bad weather is not showing the expected correlation between the risk costs and the total project cost, then this could be because the bad weather risk has been modelled as a discrete effect at each of several work sites in each of several months, hence the risk has been artificially reduced by its dispersal.

The way to measure the correlation of a single risk, distributed across several tasks in a plan or several cost elements in an estimate, is to synthesise a set of data samples from the summation of the samples of the discrete instances, correlating them against the overall total (Table 6.8).

Here, the same risk, of a snowstorm disrupting works, occurs at three separate worksites: A, B and C – see cells B2 to D2. However, it has been modelled as three separate risks and ten samples from each of these are tabled in rows 3 to 12. The correlation coefficient of this risk requires the summation of these samples as a synthetic data set, column E, for correlation with the all risks total in column F. The correlation coefficient is computed in cell I2.

Table 6.8   A synthetic set of samples

Row / column	A	B	C	D	E	F	G	H	I
2	Sample	Risk: there is a risk of a snowstorm at site A incurring a lost production cost valued at £10k to £25k.	Risk: the same snowstorm may affect site B but at a higher cost of £20k to £40k.	Risk: the same snowstorm may affect site C but at a lower cost of £5k to £10k.	Synthesised sample = sum (sample risk 1, sample risk 2 and sample risk 3)	All risks total from overall model	Correlation coefficient	= CORREL (E3:E12, F3:F12)	0.45
3	1	£16.5k	£25.0k	£9.3k	£50.8k	£753.0k			
4	2	£23.4k	£21.3k	£5.9k	£50.6k	£680.0k			
5	3	£14.7k	£33.1k	£8.6k	£56.4k	£901.0k			
6	4	£16.4k	£39.7k	£9.9k	£66.0k	£898.0k			
7	5	£11.0k	£30.5k	£8.2k	£49.7k	£920.0k			
8	6	£11.1k	£23.1k	£8.2k	£42.3k	£627.0k			
9	7	£10.6k	£23.5k	£7.4k	£41.4k	£346.0k			
10	8	£22.9k	£23.9k	£8.4k	£55.1k	£256.0k			
11	9	£13.0k	£35.5k	£8.3k	£56.7k	£734.0k			
12	10	£13.9k	£33.8k	£8.4k	£56.1k	£750.0k			

Two-dimensional Ranking

Though tornado plots (Figure 6.2) are eye-catching and have become almost a trademark of risk analysis, in reality that is about all they are. A better solution in my view is to use a far duller X-Y plot of average versus correlation coefficient (Figure 6.5).

On the basis of this ranking I would advise the PM that the largest risks are:

1. Risk 6 because of its mean value
2. Risk 1 because of its correlation coefficient
3. Risk 12 because of its mean value
4. Risk 7 because of its correlation coefficient and so on.

This insight into the largest risks is inadequate in some respects because pieces of useful information are not presented, information that three further rankings can reveal.

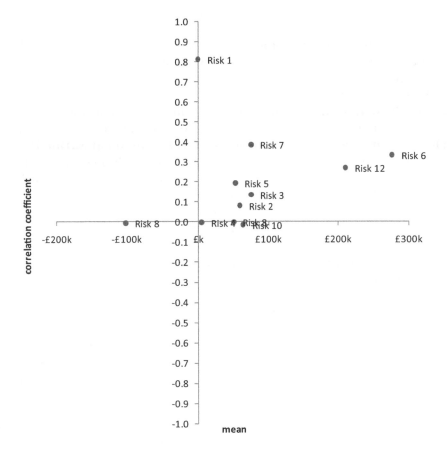

Figure 6.5   A two-dimensional risk ranking

Ranking 3: Urgency[1]

An urgency ranking places risk in an order determined by the amount of time there is left in which to do something about them. Consider the following three risks from a project to construct a new crossing to take a motorway over or under a river estuary:

1. The construction phase of the project coincides with a number of similar schemes so there may be a shortage of the expertise and plant needed to construct the tunnel.
2. The local authority may not like the proposed alignment of the tunnel because it emerges onto a greenfield site. It may prefer a brownfield site and so raise an objection at the Public Inquiry that will result in a further period of option evaluation and consultation.
3. The motorway developer is as yet undecided on a design solution. Options being considered include a single-bore tunnel, a twin-bore tunnel or a suspension bridge.
4. Say analysis of the mean risk costs of the three produces the following conclusions:

In terms of average, imagine the risks are ranked thus:

- **1 The extended construction period** - doing nothing about it could see the construction phase extended by up to two years whilst waiting for skilled resources and material supply chains to be released from other projects. This would incur a large time-dependent cost.
- **3 Unresolved design options** – the exposure to differences in the costs of the options is less than the costs of extending the construction period by up to two years.
- **2 The extended consultation precipitated by the local authority** – though it may extend the public consultation phase by an additional six months, this is a time with the lowest time-dependent cost because the project team will be at its smallest, and design and construction staff will have not yet been hired.

Although in terms of correlation the risks are ranked thus:

- **2 The extended consultation precipitated by the local authority** – this may or may not happen, and the transitions between the two scenarios are discernible in the overall model results.
- **3 Unresolved design options** – the transitions between each the risk model were discernible in the overall model results, but less so.
- **1 The extended construction period** – this will certainly happen in the modelling context as opposed to the 'may or may not' of the other two risks. Even though the size of the extension is not known, variation in it showed no strong correlation with the overall result.

The overall ranking sequences: 1-3-2 and 2-3-1, confuse as much as they illuminate and in answer to the PMs question of which risk do we tackle first – 'it all depends' – tabling either two one-dimensional tornados or one two-dimensional ranking will not help.

---

1   I am indebted to Dr Julian Downes of J & S Ltd. for this insight.

To resolve the conundrum, we need to identify the possible prevention actions that may be able to eliminate the risk and think about when these need to happen.

Risk 1, the extended construction period due to shortages of expertise and materials, may possibly be prevented by training, by bringing in experts from overseas, or by planning to avoid the dependency on the experts altogether by altering the tunnel design in some way. Likewise for materials – there may be a way to prevent the foreseen problem happening. This needs to happen before construction tenders are issued.

Risk 2, the possible delay caused by the local authority raising a formal objection at a planning inquiry, may possibly be prevented by a programme of consultation to explain away concerns, or the provision of something the authority would value – an asset or a service that would mitigate its objection. This all needs to happen before an application for permission to proceed is made.

Risk 3, the unresolved design options, may possibly be prevented by a geotechnical study that could find that there is only a single viable solution. This needs to happen before tenders for detailed design are invited.

In terms of urgency, the ranking is clear (see Figure 6.6):

- The risk prevention actions for Risk 2 need to happen first, before the preliminary design, consultation and approval phase of the project.
- These are followed by those for Risk 3, which must happen before the detailed design phase.
- And finally those for Risk 1, which must happen before the construction phase.

The ranking is 2-3-1, thought it is subject to one important sophistication.

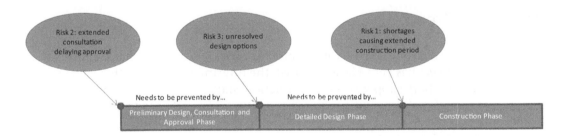

Figure 6.6 Urgency of prevention action

The sophistication is that the amount of time that could be required for the prevention to take effect needs to be taken into account, and not solely the urgency of its commencement.

This consideration could alter the ranking. In Figure 6.7, for example, the time needed (say) to make other construction resources into the procurement domain in order to avoid Risk 1 is sufficiently long, making it a more urgent risk to tackle than the unresolved design options of Risk 3. Thus the prevention action for 1 has to start before that of 2 for it to be effective. The ranking is thus 1-3-2.

For those who are asked to review projects in trouble to find out why they have failed, this concept of time-to-effect can be an important clue. The projects may have robust and

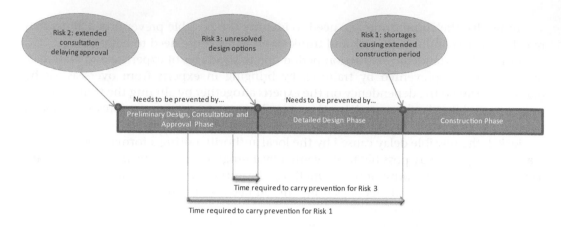

Figure 6.7   Effect of time needed for prevention to be effective

auditable risk-management processes in place and they may have teams of analysts and managers working on risk. But if the analysts have *not* ranked risks by urgency adjusted for time-to-effect then it may be that no prevention measures have time to be successful, and that the ratio of time required for prevention over the time made available for prevention, the average of which across all risks I call the Risk Treatment Effectiveness index is less than one, i.e. poor. There is a double jeopardy here: having a risk-management process in place does not guarantee that risks will be managed, and having an RTE of less than 1 ensures they never will (all) be.

Ranking 4: Trend

Consider the shortage of expertise for construction risks – Risk 1 in the previous section. Assuming the project has yet to be authorised, the unwanted outcome, that some skilled trades will be in short supply when construction starts, is some time away. Research reveals that though training in the required expertise is currently productive, the number to apprentices is due to fall. The long-term trend in the exposure associated with Risk 1 is therefore sloping downwards (Figure 6.8).

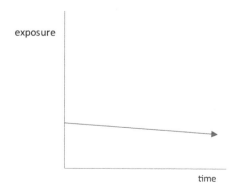

Figure 6.8   Risk 1 has a declining exposure trend

If further research on Risk 2, the local authority objecting to the apprentice scheme, finds that changes in the political make-up of the local authority may occur as a result of imminent elections. This may in turn mean that the demands made on the project for betterment and more extensive consultation could increase sharply in the near future. The exposure trend inherent in Risk 2 is moving upwards. The implication of this would be that it would be better to cut a deal with the current incumbents now than to act later. The long-term trend on this risk is therefore level, with a potential upward step (Figure 6.9).

All future changes in risk exposure are, of course, potential.

And finally, the trend in exposure associated Risk 3, unresolved design options, is constant. The exposure to the consequences of not choosing which design it will be are the same this week as next, provided we are still within the period of time in which risk prevention will be effective (see the previous section). Once that period has passed it would be a different matter, but for the moment, let's assume the exposure is constant (Figure 6.10).

On the basis of trend, the analyst should recommend Risk 2 as the most pressing risk to try to prevent, followed by 3 and finally 1: 2-3-1.

Ranking by trend needs to be done carefully. Judging exposure trends is subjective, and it is usually easy to construct an argument that it is going in the opposite direction.

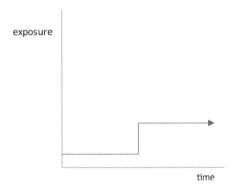

Figure 6.9   Risk 2 has a rising, stepped exposure

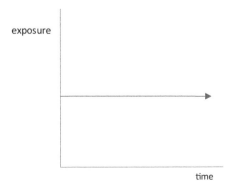

Figure 6.10 Risk 3 has a constant exposure

For example, the rising exposure of Risk 2 would be inverted if the project were to gain powers under an Act of Parliament, overruling the local authority. In such a scenario, the exposure would decrease in due course.

Or, in regard to Risk 1, shortages of skilled resources, it is possible to argue that the increased supply of skills is probably a response to a foreseen demand. Therefore the proportion of the expanding skill base available to the project will be unchanged, the exposure increasing if the project tries to hire the skills it needs at short notice in a suppliers market, consequently having to pay premium rates.

But an analyst should strongly advise against acceptance of arguments that invert the trends that depend on the project acting in ways that are contrary to the natural order of things, predicated on an presumption that it will be successful in doing so. In Risk 2, for example, ignoring the local authority would be contrary to the natural way of dealing with such an institution. The success of the strategy would be dependent on the project gaining those superior legal powers soon. Relying on such a strategy being successful is in itself a risk and I would report it as such. I would not reduce risk exposure forecasts on an assumption that against the grain heroics will be successful.

However, it is not only analysts who need to challenge managers about subjective views on risk exposure trends. Managers need to challenge analysts as well. PMs should check arguments in support of a particular trend direction that have assumed successful action by the PM will be forthcoming, just in case this simply not going to happen. Perhaps the PM does not have the powers to conclude a deal with the local authority, to influence the supply side of the skills market or to decide whether the crossing will be a bridge or a tunnel. Any of these may require action by the funders or by a government to validate the deemed trend. After all, the trend in the risk of fire during a dry spell is not downwards simply because the project can put fires out.

Ranking 5: Leverage

The Risk Reduction Leverage (RRL) is the reduction in exposure obtained, divided by the expenditure on prevention. If the reduction in exposure is £1M, and is obtained at a cost of £250,000, then the RRL is four. Risks can then be ranked according to the effectiveness with which they can be managed.

However, this neat, clear, attractively pragmatic method of ranking risks can be difficult to use in practice. This is due to the need to elicit information from experts about the possible reduction in risk exposure that could to be obtained from prevention. This compounds the already tricky problem of deciding what the risk exposure currently is – the 'before' position that the analyst presumably already has encoded in the modelling – with the need for a future position (an 'after').

There is a hidden danger too, in having before and after risk exposures featured in the risk analysis. I have also known aggressive project managers seek sufficient funding for only the 'after' position, i.e. they have assumed all risk preventions will be successful, an implication there will be no need for containment and remediation. I have seen millions of pounds wiped off risk exposures at a stroke by clients convinced of their future success at managing the risk. I have no answer to this except to say quietly that they may care to read the works of Peter Morris and Bent Flyvbjerg, to realise that the people behind those narratives were no less talented, hard-working and determined than they are.

RRL ranking can be useful in selected instances. Consider the following elaborations on the three risks used in the preceding sections.

1. There are likely to be shortages of skilled labour for construction: This risk has a mean exposure of £20M, based on the extension of time it will take to complete construction with only a limited workforce. Prevention could be effected by a training programme that would cost £5M to set up and run.
2. People living in the village do not like the proposed logistics solution and may block the road through the village. This risk has an exposure of £5M, which is the estimated additional cost of having to use the alternative, longer, route. The villagers would probably tolerate the use of the road if the village hall is refurbished at a cost of £250K.
3. The environmental assessment is not yet complete and the preferred option of a multi-arched low bridge may have to be dropped in favour of a tunnel. The tunnel solution will cost between £50M and £75M more to construct.

The prevention actions for Risks 1 and 2 are extraordinary business measures, each of which will lead to the closure of the risk. In this kind of situation, RRL works well.

For Risk 1, the RRL is £20M/£5mn = 4, and for Risk 2, the RRL is £5M/£0.25M = 20.

It is difficult to assess the RRL for Risk 3. An environmental assessment is in progress and it may determine that the more expensive tunnel option must be built. It would not be right to spend additional funds to try to influence the outcome of the assessment towards the cheaper bridge option. The concept of a cost of prevention, the denominator of the RRL equation, does not work here because the prevention action would require something tantamount to unprofessional conduct.

Generally, RRL does not work if the prevention action is unlikely to be carried out, or additionally, where it can be carried out but its cost cannot be picked out from the mix of costs that make up the day to day running expense of a project. This is a similar problem to the accounting of dividing overheads between profit centres: how does one divide the costs of ordinary business between the risk prevention costs of 100 estimating tolerance risks, for example? Here the RRL approach will not work without making approximations and assumptions about what the prevention cost per risk is, and in consequence, the RRL method will prove unsatisfactory to use.

Furthermore, the probability of the unwanted outcome of Risk 3 happening – that a tunnel is chosen and not a bridge – I suggest will remain the same no matter how much effort is spent trying to make it otherwise. The risk is beyond the control of those who may want a bridge. RRL therefore sometimes fails on a second count: the mooted prevention action will have little, or no, effect on the unwanted outcome of the risk. This is commonplace where the prevention action is essentially more management, a tighter grip, a focused attention, a rigorous application and so on. While these are certainly the right things to do, and possibly even effective ones as well, these actions do not guarantee that the cost expenditure will reduce risk directly, in the way that, say, acquiring more construction resources from overseas markets would. Their effect is indirect and RRL rankings should not be made on assumption that indirect actions will be successful in their intentions, only direct ones. If the indirect actions are not funded then, of course, the risk will almost certainly increase. However, the corollary does not hold: funding them does not ensure a decrease, and that certainty is necessary for RRL measures to be reliable as indicators of where to choose to take risk prevention action.

The ranking by RRL is therefore:

1. Risk 2: RRL = 20
2. Risk 1: RRL = 4
3. Risk 3: RRL = X (not known).

Conclusions on Ranking

The results of the five rankings: mean, correlation, urgency, trend and leverage, are in Table 6.9.

What was a simple 1-3-2 ranking by average and urgency is countered by a 2-3-1 order for correlation and trend, and 2-1 order for leverage (the leverage for Risk 3 is not calculable). So what is the answer to the question 'what are the biggest risks?'

By adding different rankings, some would rightly say that what I have done is to add complexity where none was asked for, creating confusion where clarity was desired. What I have done though is show that the answer is not as simplistic as red/amber/green and high/medium/low point-ranking schemes based from subjective opinion may have led one to assume.

My answer would not be to say 'it all depends', because that would only compound the irritation. What I would do is set out the rankings as above, in order to satisfy those who may have to report risk rankings extracted from my work, and who may have to furnish evidence of best practice to auditors and reviewers. I would then tell a story.

Risk 1 has the largest value but I recommend you defer management of that. Its prevention is not yet urgent and the trend of its exposure is not upwards. Instead, I think you ought to act to prevent Risk 2, which though it is low valued has high influence. Its exposure trend is upwards, and the time left in which to prevent it happening is reducing. Acting to prevent Risk 2 will benefit from its high reduction leverage.

Risk 3 needs to be watched. It scores mid-range on all the rankings but preventing it is not within your delegated powers. The funders will need to be made aware of any changes in its evaluation.

In summary: act on Risk 2 because it is generally ranked high and because the action should be effective. I recommend Risk 3 is watched and any changes that may alter its analysis reported back to me. Risk 1 can be ignored until next month's review.

Table 6.9   Risk rankings

Rank	Average	Correlation	Urgency	Trend	Leverage
#1	Risk 1	Risk 2	Risk 1	Risk 2	Risk 2
#2	Risk 3	Risk 3	Risk 3	Risk 3	Risk 1
#3	Risk 2	Risk 1	Risk 2	Risk 1	x

I see the question 'What are the biggest risks?' as a cipher for two very real PM-centric questions: 'What is going to hit me?' and 'How many of them are there?'

Analysts who produce a simple ranking may be able to answer the first question, though I should phrase that as an 'apparently answer', because I doubt the validity of groupthink-produced rankings that are unsupported by quantitative evidence. But without carrying out more sophisticated rankings and using these to construct a story they will not be able to answer either question adequately.

The answer is obvious: the rankings are not so much for the PM but more for the analyst. They are the means by which the analyst can construct a story, and it is the story that is for the PM.

Digressions

ATTAINING INVESTMENT QUALITY ANALYSIS

Many people who may not know much of the detail of a risk analysis will be relying on it to be valid, not only at the current time but also for time to come. This can be anything up to ten years. For example, the risk analyses that supported the construction of the Channel Tunnel Rail Link were predicting costs and timescales ten years in advance of completion. Similarly the assessments for the London Crossrail project were done in 2007 for a project predicted to complete in 2017 at an outturn cost *then* of £16B.

Risk analyses inform investment decisions and must try to be of investment quality. Clients may show loyalty to a particular risk analysis process but what matters more to them is that its outputs are reliable, complete and constant. They have a duty of care to other peoples' money, a duty that persists from the moment an investment decision is made until the time the project is completed. An analysis of investment quality is therefore one that can be relied upon to be the basis of investment decisions, now and in the future.

It is common in risk analysis to present results with a caveat that they are a work in progress and that their results reflect a snapshot in time. This may sometimes have to be the case, when one is pressured for evidence of the analysis being underway or for interim results, but for investment quality work it is necessary to affirm that the analysis is finished. Barring external events that are wilful acts detrimental to the success of the project the results of it should hold until the project is finished. Nothing more is known that could be added to increase the validity of the analysis. This may be a holy grail but it is publication of results and a 'prepared to stand by them' attitude that is the imprimatur of investment quality work. It is not the setting up of an ongoing process of generating current snapshots of a not yet completed analysis. The process of risk management should continue until the remaining risk exposure is less than the cost of managing it and the process of risk analysis should end well before that.

Controlling Changes to the Analysis

A consequence of attainment of investment quality analysis is that the numbers in it will have become almost too difficult to alter, because others have accepted them as a basis for decisions they have made. Moreover some people may consider the time for analysis has passed and so do not want to be involved in further rounds of risk identification and quantification workshops. Some may even think that changing the numbers is a sign of weakness.

However, there will probably be pressure to change them now that the risk numbers have migrated from a funding scenario to a project management situation. Numbers pertaining to costs and timescales will acquire a mutability not seen before. New risks may be identified; some risk numbers may move from being part of a model for a contingency against unwanted outcome A, to being part of a model for a contingency against unwanted outcome B; and managers will pressure analysts to present risk analyses that are nuanced towards messages of success, that demonstrate effective control, and that show there is no need for intervention from outside the project.

This is entirely understandable, but an analyst should be mindful of the needs of others who have depended on their work for their investment decisions. They need to be assured that the risk analysis remains valid as it moves into the more turbulent waters of the project management community, so I strongly recommend that it is appropriate at this time to introduce controls on the risk analysis numbers.

These should take two forms. First the numbers in the analysis – essentially the quantifications of the risks – should be approved by a committee of suitably qualified experts in the risk areas assessed. The implication of this is that analysts will need to write and publish reports that record analyses on which investment decisions have been made that the experts can review and accept.

Secondly, the numbers in the risk assessment will need to change for all sorts of valid reasons but they should only be allowed to do so in a controlled way. I strongly recommend that a scheme of delegated authority to revise risk numbers is put in place. For example, Table 6.10 shows a scheme in which the changes in risk mean cost or probability have to be approved by an increasing number of senior staff the larger they are.

I am not seeking to give analysts the sole rights to alter the numbers. That could be impractical because it may constitute both a bottleneck in the production of project information and reports, and be little more than a job creation scheme for analysts. However, I am saying that a process needs to be established once investment decisions based on risk analyses have been made. This will let the users of the risk analyses, other than its creators, adjust the risk data for their own purposes, which simultaneously keeps the analyst aware of those changes. In turn, the analyst can inform others who have relied on the risk analysis of the potential consequences. This seems sensible to me, and yet I have known only one client put it into practice.

Table 6.10   An example of a delegated authority scheme to alter risk evaluations

Mean cost change ...	.... or probability change	Must be approved by ...
$\Delta X > £100k$	$\Delta P > 25\%$	Programme director
£50k $< \Delta X <$ £99k	$10\% < \Delta P < 24\%$	Project manager
£0k $< \Delta X <$ £49k	$0\% < \Delta P < 9\%$	Design leads
£0k $< \Delta X <$ £49k	$0\% < \Delta P < 9\%$	Commercial leads

Configuration Control of the Analysis

Just as important as taking steps to stabilise the analysis from changes made at the behest of project management teams is the need to control what is often the major source of volatility in risk analyses, the analysts themselves. The greater part of the froth and bubble of risk analysis will be the ongoing development and correction of the risk models and registers. Some of this may be happening on models that have already been used to generate published results, with the possible consequences that those results cannot then be reproduced nor the analysis developed in a direction that its users want but which is different to that which was subsequently taken by the analysts. Such situations can happen if the configuration of a model is not recorded at the time its results were published. The follow-up questions users ask a week later, the what-ifs they want evaluated, and the data they want tabulated, may no longer be possible to accommodate because the analysts have pressed on with development of the model and have altered it significantly, even though it may have been in ways that are perfectly understandable: the addition of new risks, changes to the modelling of old risks, ownership revisions, corrections and so on.

What is not understandable is why the need for configuration control of the model had not been foreseen by the risk analysts. Analysts must expect to have to reproduce results from any stage of a project's development right up until the project accounts have been closed, which is typically two years after completion – even longer when claims from the contractor are disputed by the funders.

A configuration management tool would help greatly with this but the difficulty is that there is not one available that is really suitable. To my thinking this means being able to open up a model, then being able to undo step by step – <CTRL>Z if you like – every edit while a time and date stamp in a pop-up window conveniently displays a backward-running chronology of the model's development, showing a highlighted marker where the configuration tool noted that the model was run and must therefore have produced a set of results.

This does not seem to be too difficult to design, and it may even bring a bonus of shortening model development timescales. We risk analysts spend a lot of time rearranging our spreadsheet nests and correcting our encoding errors, something we might take pains to avoid if our every move were recorded. There is security and comfort in spreadsheet work. I can work alone on something that is sufficiently small-scale and obscure that I am safe from anyone looking over my shoulder, jabbing a finger at a cell and telling me that its content is wrong. Further, unlike on a production line, I do not have to hand over my work to the person next to me with the risk that they may denounce it as unusable and reject it. Reports will be read and presentations will be seen but the spreadsheets that underlie them are hardly ever scrutinised, and all in all, I am safe from the workaday, sustained, casual criticism to which other forms of work are subjected. Instead, quietly and virtually without fear of being found out, I can make and correct as many mistakes as I habitually do and yet it still looks like productive work. I was taught that when faced with behaviours that are seemingly inexplicable or inefficient, look for the economic truth because that is one that will have the ring to it, and here *perhaps*, only perhaps, here lies the reason why my configuration control tool does not exist. Correcting mistakes pays.

A Risk Model Naming Convention and Development Strategy

In the absence of a configuration management tool I would like to explain a naming convention and development strategy that goes part-way to mitigating the problem.

Risk modelling is generally performed within the context of Microsoft applications and their associated add-ins. It is a development environment at the laissez-faire end of the control scale, so much so that it is possible to earn a decent living checking completed spreadsheets for anxious clients, and while the freedoms and flexibility are certainly much to be valued, it is all too easy, even accidently, to edit a model and to forget that this has been done. Reproducibility is an important characteristic of investment quality models and so I would like to give some tips on how I go about mitigating the risk of not being able to do it.

Let me imagine a scenario in which there is a risk model that generates sets of data tables, and that these data tables are in turn used to generate results tables and graphs for inclusion in PowerPoint and Word. It almost goes without saying that every presentation and report must contain a reference to the model that generated the results, but it is also good practice for that reference to contain the file name of the model that has been augmented in a way that ensures its uniqueness. This gives it an air of authority and that tells the reader something about what the model is for. Here is an example of my formulation:

BEKO ICTRA V3 for CSR2.XLS

It is built up from four components:

1. BEKO is the project.
2. ICTRA is the type of model, in this case an integrated cost and time risk assessment. Other ones include QCRA (Quantitative cost risk assessment) and QSRA (Quantitative schedule risk assessment).
3. V3 is the version number. It acts as a reminder that there are other versions of the model that have been developed, presumably for other uses.
4. CSR2 is the name of a milestone in the development of BEKO.

The formulation is for a completed model and not one in development because it lacks the obvious inclusion of a date stamp. I find date stamps, viz:

081123 BEKO ICTRA V3 for CSR2.XLS

confuse rather than clarify when viewed, say, a year later. If (for example), the CSR2 milestone date was 090105, six weeks after the date stamp on the model, I could not be certain that I had made no further changes to the BEKO ICTRA in the intervening period between the apparent cessation of the model development and publication of the milestone reports. Did I or did I not run a what-if scenario and what happened to its results? Did my client use them instead of the originals? I do not know, and if you were to say I should have written down what I did, I would agree, but a year on my notes now seem to be unclear on this point, perhaps because the model had so many configuration switches that I did not note the status of all of them for a one-off run of the model.

And so I take the view that the date of the model is implied by the name of the milestone and its date. All results generated by this model are labelled with its name and are accompanied by a table of any data used to define the parameters and variables used by the model to generate them.

I am then left with two problems: how to accommodate post-results publication changes, and how to name the model while it is in development. I do not wish to use V4, V5 and so on because I reserve this sequence for models that are complete, their results published for a specific purpose:

1. BEKO ICTRA V1 for Feasibility Report.XLS
2. BEKO ICTRA V2 for Financial Submission.XLS
3. BEKO ICTRA V3 for CSR2.XLS
4. BEKO ICTRA V4 for Financial Close.XLS
5. BEKO ICTRA V5 for Contract Award.XLS

I am equally wary of giving a model in development a name that gives it a spurious authority that may be misunderstood by users of any interim results, to which its imprimatur is attached:

BEKO ICTRA V5 for Contract Award draft A.XLS

Instead I prefer to choose a name for models in development which clearly communicates that they are temporary in some unspecified way. Thus, I use the following sequence:

1. AAA.XLS
2. BBB.XLS
3. CCC.XLS ... and so on

This is the style of model name that appears on all results I hand over to the client for whatever purpose they intend. I then ask them that if these results are to be published internally or externally, potentially having to be reproduced, could I please be told so that I can rename the file according to my convention, thus:

DDD.XLS

becomes:

BEKO ICTRA V5 for the 2011 Annual Construction Report.XLS.

At no point during the development of the model do I issue results with anything other than a junk label of the AAA form. I only apply a formal name if the client specifically requests it or if I know the results will be published by the client. When clients come back a year or so later with a graph labelled TTT and ask did it or did it not contain X? I have to point out that I may not be able to find out because TTT was a model in a rolling development sequence, and had I been told of the need to preserve its results, I would have given the underlying model a formal name.

Once a model has been formally named, I then open a new folder with the next version number as its title. Often, if I am not yet aware of what it will be, I do this without the milestone part, and into this I copy the last formally named model which I re-title AAA, and onwards the development continues.

- BEKO ICTRA V5
  1. AAA
  2. BBB
  3. CCC
  4. DDD
  5. EEE
  6. ~~EEE~~ / BEKO ICTRA V5 ICTRA for Annual Construction Report.XLS 2011.XLS.

- BEKO ICTRA V6
  1. BEKO ICTRA V5 ICTRA for Annual Construction Report.XLS/ AAA
  2. AAA
  3. BBB

The commonplace scenario of a client wanting a 'what-if' evaluated on BEKO ICTRA V5 ICTRA for Annual Construction Report.XLS 2011.XLS after it has been published is easy to do without corrupting this formally named, and hence, frozen, file by using AAA from the subsequent on BEKO ICTRA V6 directory.

A Working Practice for Handling Risk Model Results

Modelling runs of RMT can be set up to produce automatically time- and date-stamped worksheets of samples and statistics. It is possible to include these in the same workbook as the model itself, but I do not because it can increase the size of the model to a level at which opening, saving and closing it can become excessively protracted. Instead I keep the output results sheets in files whose titles closely match the model used to generate them. These workbooks I call 'wraps' after the movie director call:

BEKO ICTRA V3 for CSR2 wrap 1.XLS

BEKO ICTRA V3 for CSR2 wrap 2.XLS

AAA wrap 1.XLS and so on.

Different wraps from the same model reflect different model set-ups; different tagging, Boolean conditions on joint outputs, control switches switched off or on, parameters changed etc. It is good practice to encode the set up of the model in a front worksheet so that this can be copied to a wrap.

Wraps contain just the raw output – the numerical data – from a model. Any graphs and output data tables needed for presentations and reports I then create using references to these data. I do this directly, or sometimes after including some desired post-modelling analysis such as rankings, statistics computations and so forth. I keep these calculations, graphs and tables in additional sheets in the wrap workbooks. Nothing that is intended for use in reports or presentations is ever created outside of the wraps. I never create results using Word or in PowerPoint because I cannot easily keep the working out alongside them nor can I trace back a graph, say, to its raw data output. Instead every figure seen or table read has its master copy in a wrap. Indeed I never even do the colouring, bordering and highlighting outside of a wrap because when I have done, I will have altered it from the plain form that by implication exists in the wraps and I have found that rather than reinforce a memory of what the table or figure looks like, this weakens it, sometimes to the extent that I have convinced myself I have lost primary results when I have not: they were just listed in a different order or coloured differently from the originals in the wraps. As my colleague J observed, when you are looking for numerical data in your records, it is not the numbers you look for first but the image of them that you have in your mind.

The advantage of using a wrap is that the raw output data that produced a set of results is there forever if needs be and can be reused to produce further figures, while development of the model: AAA, BBB, CCC etc. can proceed: AAA wrap 1 is preserved.

If a client wants a different set of results generated from a wrap to that previously supplied, and it is not desirable that they are simply incorporated as additional sheets (perhaps because the existing wrap workbook contains privileged information but the additions have to be published externally), I follow the naming convention adopted for the model:

1. AAA wrap 1 stage gate 3 statistics.XLS
2. AAA wrap 1 contractor exposure graphs.XLS
3. AAA wrap 1 monthly report trend tables.XLS and so on.

All of these contain the same raw output data. If this data would not let me compute the results wanted, perhaps because I had not foreseen the need to capture certain statistics, I needed to have set up the model differently and then I would reflect this in the number of the wrap. If each of the above wraps came from a different modelling run of the same model but with different configurations, the sequence would be:

1. AAA wrap 1 stage gate 3 statistics.XLS
2. AAA wrap 2 contractor exposure graphs.XLS
3. AAA wrap 3 monthly report trend tables.XLS

and if they came from the same model but from different stages of its development, then (for example)

1. AAA wrap 1 stage gate 3 statistics.XLS
2. DDD wrap 1 contractor exposure graphs.XLS
3. BEKO V3 for CSR2 wrap 1 monthly report trend tables.XLS and so on.

Not all requests for additional figures and tables that apparently call for a new set-up of the model, a re-run and a new wrap need this to be done. Sometimes the new results can be synthesised from the old ones in an existing wrap. For example if the model has 6 risks, and is run for 1000 iterations, then if the wrap contains the 6 × 1000 data samples, all permutations of risks can be subsequently synthesised within the wrap. Say for example one risk has been closed then, as shown in Table 6.11, by accumulating the samples for the other five risks into a concurrent total column (the right-most one), a data set can be synthesised that will reflect the statistics of the six versus the five, hence showing the impact of the removal of four without recourse to switching the risk off in the model and re-running it to create a wrap specific to the scenario.

The same thing can be done with any number of subtractions in order to synthesise subsets of raw data for subsequent analysis. Further, I could create a risk model with one risk in it, run this in the same modelling conditions as the main model for the same number of samples, paste its raw data alongside the wrap raw data and assess the potential impact of adding a new risk to the model.

Wraps open up all manner of possibilities, while crucially keeping errant fingers and absent minds away from the model itself. Imagine the feeling if the potential impact of a risk had been assessed by editing out its modelling so that a set of 'without' statistics

Table 6.11   Synthesising a data subset in a wrap

Sample number	Risk 1	Risk 2	Risk 3	Risk 4	Risk 5	Risk 6	Total risks 1 to 6	Total, less risk 4
1	£20.6k	£25.0k	£16.5k	£25.0k	£9.3k	£103.0k	£199.5k	£174.5k
2	£21.2k	£20.0k	£23.4k	£21.3k	£5.9k	£106.2k	£198.0k	£176.7k
3	£20.1k	£25.0k	£14.7k	£33.1k	£8.6k	£100.6k	£202.1k	£169.0k
4	£19.0k	£50.0k	£16.4k	£39.7k	£9.9k	£94.8k	£229.8k	£190.1k
5	£18.9k	£50.0k	£11.0k	£30.5k	£8.2k	£94.5k	£213.1k	£182.6k
6	£19.7k	£50.0k	£11.1k	£23.1k	£8.2k	£98.4k	£210.4k	£187.4k
7	£20.9k	£50.0k	£10.6k	£23.5k	£7.4k	£104.6k	£216.9k	£193.4k
8	£19.5k	£25.0k	£22.9k	£23.9k	£8.4k	£97.4k	£197.0k	£173.2k
9	£20.4k	£37.5k	£13.0k	£35.5k	£8.3k	£101.8k	£216.4k	£180.9k
10	£19.5k	£50.0k	£13.9k	£33.8k	£8.4k	£97.7k	£223.4k	£189.5k
–	–	–	–	–	–	–	–	–
989	£21.4k	£50.0k	£12.1k	£38.4k	£7.9k	£107.1k	£236.8k	£198.4k
990	£20.4k	£25.0k	£12.2k	£34.7k	£9.8k	£101.9k	£204.0k	£169.3k
991	£19.3k	£25.0k	£22.0k	£31.0k	£7.4k	£96.4k	£201.0k	£170.1k
992	£19.8k	£25.0k	£24.5k	£37.1k	£6.9k	£99.0k	£212.2k	£175.1k
993	£20.2k	£37.5k	£24.2k	£29.1k	£6.2k	£100.9k	£217.9k	£188.9k
994	£20.4k	£25.0k	£17.5k	£36.3k	£5.9k	£102.2k	£207.4k	£171.1k
995	£19.9k	£50.0k	£24.6k	£31.4k	£7.3k	£99.3k	£232.4k	£201.0k
996	£19.6k	£37.5k	£19.9k	£21.8k	£8.1k	£98.1k	£205.0k	£183.2k
997	£19.3k	£50.0k	£14.2k	£29.6k	£6.0k	£96.3k	£215.4k	£185.8k
998	£20.1k	£37.5k	£13.2k	£35.8k	£6.6k	£100.4k	£213.7k	£177.8k
999	£20.9k	£50.0k	£21.0k	£23.7k	£5.3k	£104.3k	£225.1k	£201.4k
1000	£20.1k	£37.5k	£24.1k	£30.8k	£9.5k	£100.3k	£222.3k	£191.5k
						Mean value	£212.5k	£182.5k

could be generated for comparison alongside a set of 'with'. Then discovering a year or so later that for some reason the risk's modelling had never been reinstated afterwards.

Completing the above may be a wrap of the analysis but it is not necessarily a wrap for the analyst. Completing a set of reports and presentations certainly presents an opportunity for the analyst to depart the scene, leaving the PM to the slings and arrows of outrageous misfortune. But there is something more an analyst can do to help the project manager that will benefit everyone involved with the project.

SOLVING THE RISK MODEL TO DERIVE THE RISK REDUCTION STRATEGY

The world of project management is populated by many who seem to those supplying goods and services to them to lack an appreciation of value, and are instead fixated simply on cost. They relentlessly apply a maxim that more cost is bad and less is better, though only for a short while until even less expenditure becomes the objective. It is the same with time: more is bad and less is good. This is a role most project managers I know keep up like method actors, who stay in character at all times while working. I believe it is not an inherent behaviour, but one that has been acquired by the funders habitually giving the PMs slightly less funding than they will need to deliver the project. Do it for less. Do it quicker.

So if an analyst presents a PM with an exposure curve that has been carefully computed to fit the PM's risks, and a comprehensive and well thought-out ranking that is accompanied by a story of how the risks could be controlled, what usually happens? The PM makes a mark on the x axis (see Figure 6.11) and says something to the effect of 'That's my budget. Having you tell me there is a large probability I will overspend doesn't alter things. That's all I've got, and all I am going to get. If I don't deliver it for that my career will flounder, and so I'm going to drive every cost down and down, starting with yours.'

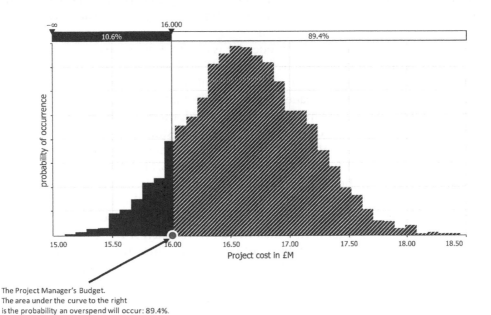

The Project Manager's Budget.
The area under the curve to the right
is the probability an overspend will occur: 89.4%.

Figure 6.11 The project manager's problem

Should the analyst break out into a sweat? I used to. Diligent work is dismissed at the very moment analysts hope they will be receiving a modicum of appreciation and thanks. However, analysts can respond with an offer no rational PM could refuse. They can show how the project could be delivered on time and to budget and what is more, prove it. Analysts can derive the risk management solution a project would most benefit from. Moreover this solution will not be found to be nothing more than a strident call for more what's good – grip, commitment, drive, focus, rigour, attention etc. – and less of what's bad – waste, delay, diversion, silo working, poor communication, protective behaviours etc.

A risk management solution can be found as follows: the cost function of Figure 6.11 is the output of a set of equations. The PM's set of equations have to be solved, resulting in an outcome to the left of the budget line on the figure. This isn't probable, but neither is it impossible. There are an infinite number of points to the left of the line and so there will be just as many solutions to the set of equations. But they will all have two characteristics: each will be an acceptable outcome because the project will be delivered under budget; and each will require a possibly different combination of risk eliminations, risk reductions and risk acceptances (i.e. risks that will be left unchanged) in order to happen.

If a combination can be found that is practical to implement then that will be a risk management plan for the project. It is at such a moment, perhaps the only one in the business of risk analysis, that the hard-boiled PM may see the value of the risk analyst's work and not just its cost.

Only two actions can be taken against a risk: elimination or limitation. When solving a risk model to provide a risk management solution (which is essentially numeric) for working up into a risk management plan (which is essentially textual) the analyst must find a permutation of risk eliminations and limitations. By using the term 'permutation' and not 'combination', I am admitting the possibility that only some risks, and not absolutely all of them, will need elimination or limitation in order to achieve the desired result, an outturn cost to the left of the line.

After running the example risk model for (say) 5000 iterations, if the collection of output data samples has been switched on, a RMT tool will have been able to generate a large table of risk samples that with rearrangement of the columns and the labelling of the column headings could be made to look like the Table 6.12.

The columns contain samples of 11 risks, and the right hand column is their summation added to the fixed spot cost, the project estimate. Risk 11 in the example model is not modelled and thus not shown. The probability density function of the twelfth column is the cost function and so, from Figure 6.11, it must contain totals that are equal to or less than the PM's budget. If the whole table is sorted on the values in the twelfth column using the standard Excel tool available within the Data tab or ribbon, it is possible to scan down the sorted right-most column until a point is reached where the project manager's budget is exceeded (here I have taken it to be £16M, as in Figure 6.11). This change is marked in Table 6.13 by a dotted line. Totals less than £16M are highlighted and are above the dotted line, and those over are greyed-out and below it.

All of the risk data values in the columns to the left of the total and above the line represent risk outcome scenarios that, if they could be brought about, would give an outturn project cost less than the budget. Those below the lines are risk outcome scenarios

Table 6.12 Data samples for risks and their total inclusive of the estimate

Iteration	Risk 1 — P way spread on Q	Risk 2 — Cabinets spread on Q and R	Risk 3 — Track crossing for cables: unresolved option	Risk 4 — Compensation for late completions	Risk 5 — Land take unresolved option	Risk 6 — Emerging cost chain for bridge replacement	Risk 7 — Emerging cost chain for embankment slip	Risk 8 — Productivity improvements	Risk 9 — Public harm risk	Risk 10 — Delayed access	Risk 12 — PM charges are time dependent	Total cost (including the estimate)
1	£294k	£91k	£0k	£0k	£17k	£334k	£0k	−£100k	£50k	£59k	£165k	£16.8M
2	−£404k	£65k	£0k	£5k	£0k	£407k	£0k	−£100k	£50k	£72k	£230k	£16.2M
3	£667k	£36k	£150k	£5k	£0k	£366k	£0k	−£90k	£50k	£64k	£178k	£17.3M
4	£220k	£35k	£150k	£0k	£0k	£350k	£268k	−£120k	£50k	£64k	£247k	£17.1M
5	£266k	£110k	£150k	£8k	£0k	£360k	£0k	−£110k	£50k	£61k	£154k	£16.9M
6	−£489k	£88k	£0k	£0k	£0k	£387k	£0k	−£100k	£50k	£73k	£280k	£16.2M
7	£12k	£36k	£0k	£10k	£0k	£368k	£0k	−£100k	£50k	£61k	£281k	£16.6M
8	−£270k	£104k	£150k	£10k	£0k	£0k	£0k	−£100k	£50k	£63k	£322k	£16.2M
9	£175k	£62k	£0k	£8k	£0k	£410k	£0k	−£90k	£50k	£61k	£195k	£16.7M
10	−£298k	£86k	£0k	£3k	£0k	£327k	£0k	−£110k	£50k	£70k	£202k	£16.2M
↓	↓	↓	↓	↓	↓	↓	↓	↓	↓	↓	↓	↓
4991	£221k	£35k	£150k	£3k	£45k	£350k	£0k	−£110k	£50k	£56k	£198k	£16.9M
4992	−£158k	£0k	£0k	£8k	£0k	£372k	£0k	−£110k	£50k	£70k	£96k	£16.2M
4993	£8k	£103k	£0k	£8k	£394k	£423k	£335k	−£100k	£50k	£66k	£293k	£17.5M
4994	£63k	£64k	£150k	£3k	£0k	£378k	£0k	−£100k	£50k	£63k	£164k	£16.7M
4995	−£752k	£36k	£0k	£10k	£0k	£0k	£0k	−£100k	£50k	£71k	£189k	£15.4M
4996	−£163k	£104k	£0k	£5k	£0k	£385k	£0k	−£110k	£50k	£58k	£315k	£16.5M
4997	£356k	£92k	£0k	£5k	£0k	£357k	£0k	−£90k	£50k	£68k	£185k	£16.9M
4998	£700k	£34k	£150k	£8k	£0k	£333k	£0k	−£100k	£50k	£71k	£119k	£17.2M
4999	−£674k	£65k	£0k	£0k	£0k	£333k	£0k	−£100k	£50k	£58k	£267k	£15.9M
5000	−£55k	£35k	£150k	£5k	£0k	£398k	£0k	−£90k	£50k	£64k	£237k	£16.7M

Table 6.13 Sorted data samples

Iteration	Risk 1 P way spread on Q	Risk 2 Cabinets spread on Q and R	Risk 3 Track crossing for cables: unresolved option	Risk 4 Compensation for late completions	Risk 5 Land take unresolved option	Risk 6 Emerging cost chain for bridge replacement	Risk 7 Emerging cost chain for embankment slip	Risk 8 Productivity improvements	Risk 9 Public harm risk	Risk 10 Delayed access	Risk 12 PM charges are time dependent	Total Total cost (including the estimate)
2917	–£974k	£0k	£150k	£5k	£0k	£0k	£0k	–£100k	£50k	£64k	£59k	£15.1M
4776	–£887k	£91k	£0k	£5k	£0k	£0k	£0k	–£110k	£50k	£62k	£51k	£15.1M
2521	–£964k	£0k	£0k	£5k	£0k	£0k	£0k	–£100k	£50k	£73k	£214k	£15.2M
3436	–£864k	£34k	£0k	£5k	£0k	£0k	£0k	–£100k	£50k	£64k	£114k	£15.2M
1543	–£894k	£0k	£150k	£0k	£0k	£0k	£0k	–£110k	£50k	£58k	£54k	£15.2M
↓	↓	↓	↓	↓	↓	↓	↓	↓	↓	↓	↓	↓
2478	–£187k	£0k	£150k	£5k	£0k	£0k	£0k	–£110k	£50k	£59k	£109k	£15.9M
3562	£728k	£89k	£150k	£3k	£201k	£347k	£0k	–£100k	£50k	£60k	£6k	£16.0M
2860	–£687k	£36k	£0k	£8k	£197k	£350k	£0k	–£90k	£50k	£63k	£152k	£16.0M
↓	↓	↓	↓	↓	↓	↓	↓	↓	↓	↓	↓	↓
4126	£818k	£65k	£0k	£3k	£0k	£355k	£888k	–£90k	£50k	£58k	£478k	£18.5M
3415	£717k	£35k	£150k	£5k	£0k	£337k	£1080k	–£110k	£50k	£68k	£316k	£18.5M

that would cause an overspend to occur, so these can be ignored because the PM would not be interested in bringing about such situations.

Within the risk scenarios of interest, some risks will have zero values and some finite values, both positive and negative. The zero values represent desirable risk eliminations, and the positive finite values desirable risk limitations. Negative risk values represent realised opportunities. Each row above the dotted line in Table 6.13 is therefore a potential risk management solution.

There could be many such scenarios, it all depends on how many iterations of the model were run and how far to the right the budget line sits on the spread cost curve. The more iterations of the model there are, then the more potential solutions will be generated. The further to the right the PM's budget line sits on the cost function, the more potential solutions there will be.

Which one to offer the project manager is for the analyst to decide. My strategy for choosing one solution to recommend is to compute the average value, and the correlation coefficient with the total risk, for each discrete risk and to place these values at the top of the sorted and selected risk data (Table 6.14).

I then sort the columns from left to right (Excel will do this) in order of descending risk mean (Table 6.15).

Table 6.14 Sorted, selected samples with all sample means and correlation coefficients

Iteration	Risk 1 P way spread on Q	Risk 2 Cabinets spread on Q and R	Risk 3 Track crossing for cables: unresolved option	Risk 4 Compensation for late completions	Risk 5 Land take unresolved option	Risk 6 Emerging cost chain for bridge replacement	Risk 7 Emerging cost chain for embankment slip	Risk 8 Productivity improvements	Risk 9 Public harm risk	Risk 10 Delayed access	Risk 12 PM charges are time dependent	Total cost (including the estimate)
Mean	£0k	£59k	£75k	£5k	£53k	£277k	£75k	-£102k	£50k	£63k	£211k	
correlation coefficient	0.81	0.08	0.14	0.00	0.19	0.33	0.39	-0.01		-0.01	0.27	
Iteration no.												
2917	-£974k	£0k	£150k	£5k	£0k	£0k	£0k	-£100k	£50k	£64k	£59k	£15.1M
4776	-£887k	£91k	£0k	£5k	£0k	£0k	£0k	-£110k	£50k	£62k	£51k	£15.1M
2521	-£964k	£0k	£0k	£5k	£0k	£0k	£0k	-£100k	£50k	£73k	£214k	£15.2M
3436	-£864k	£34k	£0k	£5k	£0k	£0k	£0k	-£100k	£50k	£64k	£114k	£15.2M
1543	-£894k	£0k	£150k	£0k	£0k	£0k	£0k	-£110k	£50k	£58k	£54k	£15.2M
→	→	→	→	→	→	→	→	→	→	→	→	→
2478	-£187k	£0k	£150k	£5k	£0k	£0k	£0k	-£110k	£50k	£59k	£109k	£15.9M
3562	-£728k	£89k	£150k	£3k	£201k	£347k	£0k	-£100k	£50k	£60k	£6k	£16.0M
2860	-£687k	£36k	£0k	£8k	£197k	£350k	£0k	-£90k	£50k	£63k	£152k	£16.0M
→	→	→	→	→	→	→	→	→	→	→	→	→
4126	£818k	£65k	£0k	£3k	£0k	£355k	£888k	-£90k	£50k	£58k	£478k	£18.5M
3415	£717k	£35k	£150k	£5k	£0k	£337k	£1080k	-£110k	£50k	£68k	£316k	£18.5M

Table 6.15 Columnar sort based on the mean, with solutions >£16m removed

Iteration	Risk 6 Emerging cost chain for bridge replacement	Risk 12 PM charges are time dependent	Risk 3 Track crossing for cables: unresolved option	Risk 7 Emerging cost chain for embankment slip	Risk 10 Delayed access	Risk 2 Cabinets spread on Q and R	Risk 5 Land take unresolved option	Risk 9 Public harm risk	Risk 4 Compensation for late completions	Risk 1 P way spread on Q	Risk 8 Productivity improvements	Total cost (including the estimate)
Mean	£277k	£211k	£75k	£75k	£63k	£59k	£53k	£50k	£5k	£0k	-£102k	→
Correlation coefficient	0.33	0.27	0.14	0.39	-0.01	0.08	0.19	#DIV/0!	0.00	0.81	-0.01	
2917	£0k	£59k	£150k	£0k	£64k	£0k	£0k	£50k	£5k	-£974k	-£100k	£15.1M
4776	£0k	£51k	£0k	£0k	£62k	£91k	£0k	£50k	£5k	-£887k	-£110k	£15.1M
2521	£0k	£214k	£0k	£0k	£73k	£0k	£0k	£50k	£5k	-£964k	-£100k	£15.2M
3436	£0k	£114k	£0k	£0k	£64k	£34k	£0k	£50k	£5k	-£864k	-£100k	£15.2M
→	→	→	→	→	→	→	→	→	→	→	→	→
4192	£340k	£254k	£0k	£0k	£71k	£66k	£0k	£50k	£5k	-£601k	-£110k	£15.9M
357	£381k	£228k	£0k	£0k	£67k	£36k	£0k	£50k	£0k	-£567k	-£120k	£15.9M
1954	£0k	£211k	£0k	£0k	£59k	£0k	£0k	£50k	£3k	-£177k	-£70k	£15.9M
1779	£373k	£255k	£0k	£0k	£57k	£65k	£0k	£50k	£8k	-£631k	-£100k	£15.9M
2478	£0k	£109k	£150k	£0k	£59k	£0k	£0k	£50k	£5k	-£187k	-£110k	£15.9M

Each row: iteration 4778, iteration 2621 and so on, produces a set of risk outcomes that jointly give a project cost within the budget of £16M. I can choose one – any one – and see if I can construct a plan for managing the risks so that their outcomes should be as quantified in the iteration. Take iteration 4192 (highlighted) from the Table 6.15 (Table 6.16):

Table 6.16   A solution

#4192	Risk	Mean	Correlation coefficient	Desired outcome
Risk 6	Emerging cost chain for bridge replacement	£277k	0.33	£340k
Risk 12	PM charges are time-dependent	£211k	0.27	£254k
Risk 3	Track crossing for cables: unresolved option	£75k	0.14	£0k
Risk 7	Emerging cost chain for embankment slip	£75k	0.39	£0k
Risk 10	Delayed access	£63k	−0.01	£71k
Risk 2	Cabinets spread on Q and R	£59k	0.08	£66k
Risk 5	Land take unresolved option	£53k	0.19	£0k
Risk 9	Public harm risk	£50k		£50k
Risk 4	Compensation for late completions	£5k	0.00	£5k
Risk 1	P way spread on Q	£0k	0.81	−£601k
Risk 8	Productivity improvements	−£102k	−0.01	−£110k
	Total cost (including the estimate)			£15948k

Let me speculate on how I might present this as a solution to the project manager. Look at the entries for Risks 1 and 8 at the bottom of the table. Risk 1 is an estimating tolerance so has both exposure and opportunity characteristics. Risk 8 is an opportunity that productivity rates will be higher than allowed when the estimate was prepared (see the example model for details of these). For a solution based on #4192 to work there must be reductions in the quantity of rail required so that the outcome of Risk 1 is £601k less than the estimate, i.e. an opportunity of that value is found and then realised. In addition to this, productivity improvements worth £110k, the desired outcome of Risk 8 (an opportunity), will also have to be realised for the solution to be viable.

If the PM thinks these desired outcomes cannot be realised then a solution based on #4192 is not going to work. I will need to return to Table 6.15 and look for a solution that may be more practicable.

Which one it is could be indicated by the answer to the following question: if the opportunity in Risk 1 of £601k cannot be realised, then how much could be?

Say the answer is by £300k. Then iterations #1954 and #2478 from Table 6.16 both qualify as they require the opportunity in Risk 1 to have outcomes of less than £300k: savings of £177k and £187k respectively. Let's say we choose #1954 and tabulate it on its own (Table 6.17).

Table 6.17 A second solution

#1954	Risk	Mean	Correlation coefficient	Desired outcome
Risk 6	Emerging cost chain for bridge replacement	£277k	0.33	£0k
Risk 12	PM charges are time-dependent	£211k	0.27	£211k
Risk 3	Track crossing for cables: unresolved option	£75k	0.14	£0k
Risk 7	Emerging cost chain for embankment slip	£75k	0.39	£0k
Risk 10	Delayed access	£63k	−0.01	£59k
Risk 2	Cabinets spread on Q and R	£59k	0.08	£0k
Risk 5	Land take unresolved option	£53k	0.19	£0k
Risk 9	Public harm risk	£50k		£50k
Risk 4	Compensation for late completions	£5k	0.00	£3k
Risk 1	P way spread on Q	£0k	0.81	−£177k
Risk 8	Productivity improvements	−£102k	−0.01	−£70k
	Total cost (including the estimate)			£15949k

Before the #1954 is appraised, we can bear in mind that the PM thinks £300k can be saved on the cost of the rail required. But since the prospective solution needs only £177k, we have £123k surplus to cover other risks.

The analyst now needs to discuss with the PM the desired outcome of each risk in Table 6.17 to see if there is a way in which it can be brought about.

- Risk 6: emerging cost chain for bridge replacement

The desired outcome is £0, which implies Risk 6 needs to be eliminated. The exposure arose because blocked drainages may have allowed hydrostatic pressure to build behind the bridge abutments. It may be possible that the drains could be repaired under the existing railway maintenance contract so that the pressure would then be relieved, avoiding the risk of the abutments moving. If the repairs can be arranged then Risk 6 can be eliminated.

- Risk 12: PM charges are time-dependent.

Let us imagine that the PM is uncomfortable with the idea the project could finish early, and wants to keep the £211k of funding this solution will give them to cover the time-dependent costs that would arise if the project ran late. This is a typical response in my experience and one that I would support. I cannot recall a project that has not had its completion day imposed, unlike the estimate, and consequently has struggled to find a way to deliver it within the time allocated.

- Risk 3: track crossings for cable: unresolved option.

Here, the risk is that a basic plastic ducting will have to be rejected in favour of deeper-set, concrete piping. This design solution is usually chosen where the cables to be placed in the duct are safety-critical and so must not to be subject to flexing. The desired outcome from Table 6.17 is £0, which means that the risk has to be eliminated for the solution to work.

The analyst and the PM may research where the safety-critical cables currently run – after all, the railway currently exists and operates. It may be realised that the existing infrastructure for safety-critical cables is fit for purpose and can be reused. If so, the risk can be eliminated and hence the £0 outcome realised.

A risk management solution based on #1954 is beginning to look increasingly possible on the basis of the risks appraised so far. However, the next risk presents a problem

- Risk 7: emerging costs of embankment slip.

Here the desired outcome from Table 6.17 is £0, which means the risk has to be eliminated for the solution to work. This is not really possible because whether or not an embankment slip will occur is an external event, an act of nature, not under the control of the project team. As a result, some funding to deal with the possible costs of containment and remediation will be needed.

The £123k of surplus created by the PM being able to realise £300k worth of opportunity on Risk 1, the quantity of rail required, when only £177k was needed for the #1954 solution, could be allocated to Risk 7. The idea looks promising because the mean exposure of Risk 7 is £75k and so is amply covered by the £123k. But there is a danger in the detail. The exposure of Risk 7 only is shown in Figure 6.12.

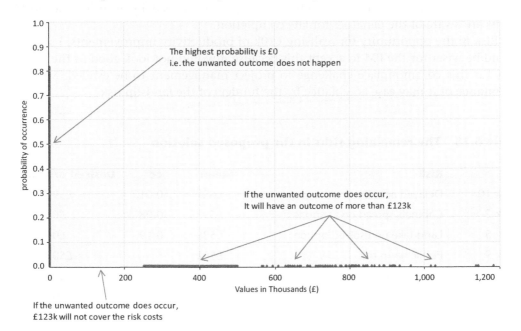

Figure 6.12 The exposure of Risk 7 only

There is a spike at £0 that represents the situation that the unwanted outcome does not occur. If it does, then any of the values associated with the unwanted outcomes will require more than £123k of funding. These are shown by a slew of low-probability, high-cost marks arrayed along the X axis, all of which are greater than £123K. Allocating the surplus £123k to the risk would therefore be futile. The average exposure of Risk 7, £75k, is caused by the relatively high probability that the unwanted outcome will not occur at all.

The advice I would give in such a situation would be to take the risk and keep the £123k for something else. I would also advise the funders are told of this so that they can, if they wish to, set aside a larger reserve from their own reserves, say £250–£450k, based on the above figure. The analyst should always try to ensure the funders are informed, possibly by offering to write a paper on behalf of the project manager recommending them to make the appropriate contingency.

I hope I have written enough about the idea of solving risk models for the last few risks in # 1954 to be dealt with in a block (Table 6.18).

Risk 10: delayed access needs to have its exposure limited to £59k. This can probably be achieved by the PM team trying to be more successful at securing access so that the statistics of the historical delays are improved.

Risk 2: uncertainty in the number of equipment cabinets needs to be eliminated so that its exposure is £0. This can probably be done by transferring it under a design and build contract to the equipment supplier.

Risk 5: unresolved land acquisition needs to be eliminated. The spare £123k should probably be used for this risk.

Risk 9: public harm. It was a fixed reserve in the model and is in the solution. No action need be taken.

Risk 4: compensation for late completions needs to have its exposure limited to £3k. This can probably be achieved by better working practices, say by ensuring the working teams are aware of the penalties for late completion.

Risk 8: the opportunity for realising £70k of productivity improvements. I suggest it would be wiser for the PM to accept this because it would not look good in the funders' eyes for this commonplace challenge to project management to be evaded. Moreover, acceptance of it may ease acceptance by the funders of the landslip risk.

Table 6.18   The remaining risks in the proposed solution

	Risk	Mean	cc	Desired outcome
Risk 10	Delayed access	£63k	−0.01	£59k
Risk 2	Cabinets spread on Q and R	£59k	0.08	£0k
Risk 5	Land take unresolved option	£53k	0.19	£0k
Risk 9	Public harm risk	£50k		£50k
Risk 4	Compensation for late completions	£5k	0.00	£3k
Risk 8	Productivity improvements	−£102k	−0.01	−£70k

The overall solution looks like this (Table 6.19).

Table 6.19 The solution

#1954	Risk	Mean	Correlation coefficient	Desired outcome	The risk management action
Risk 6	Emerging cost chain for bridge replacement	£277k	0.33	£0k	Eliminate the risk by repairing the drains under the existing maintenance contracts so that pressure on the abutments is relieved so that they do not move inwards when the deck is removed.
Risk 12	PM charges are time-dependent	£211k	0.27	£211k	Accept.
Risk 3	Track crossing for cables: unresolved option	£75k	0.14	£0k	Eliminate the risk by using the existing track crossing infrastructure.
Risk 7	Emerging cost chain for embankment slip	£75k	0.39	£0k	Accept.
Risk 10	Delayed access	£63k	−0.01	£59k	Improve management practice in order to limit exposure.
Risk 2	Cabinets spread on Q and R	£59k	0.08	£0k	Eliminate the risk by transferring it to a design and build contract.
Risk 5	Land take unresolved option	£53k	0.19	£0k	Cover with £123k accrued from risk 1, quantity of railway.
Risk 9	Public harm risk	£50k		£50k	Retain this fixed provision.
Risk 4	Compensation for late completions	£5k	0.00	£3k	Improve on-site working practice in order to limit exposure.
Risk 1	railway spread on Q	£0k	0.81	−£177k	Realise £300k of opportunity and use the excess to cover risk 5, land take.
Risk 8	Productivity improvements	−£102k	−0.01	−£70k	Accept the need to deliver 70k of productivity improvements.
	Total cost (including the estimate)			£15949k	

In the far-right column of Table 6.19 is a risk management plan that does not rely on a blind, simplistic and possibly doomed to fail attempt to manage the risks by tackling the top risks. It is analysis-led, evidence-based, management-consulted and intelligently planned. I think it is one that has an innately better chance of successfully controlling the risk.

There are two final points:

1. First, the sums of money that make up the solution can be moved from risk to risk along the solution row, adding and subtracting them as necessary. Exact fidelity to the individual values of the solution is not essential, provided the overall total does not change. A particular iteration, like #1954, is therefore best thought of as a trial solution, analysis and discussion of which should lead to an agreed solution.
2. If, in spite of good endeavours on everyone's part to control the risk, it looks like a live project is going over budget, the analyst should amend the risk model with the actual outcomes of the risks, re-solving the model to see if a solution for delivery within the remaining budget can still be found. If one can't be found, the analyst should be able to predict the extent of the forthcoming overspend. Does this mean the analyst got it wrong in the first place? Yes, it does.

A Method for Identifying Risks

I have stuck to an approximately chronological sequence of stages for risk work so far. It may seem strange that after so much analysis it is only now that I arrive at the seminal matter of risk identification. Even I am curious as to why, but I think the reason is this: having informed the funders and helped the project manager, it is around about now that the first wave of contractors tend to arrive to start the work of detailed design and construction, and one of the ever so slightly ceremonial tasks that befalls the analyst at this time is to conduct risk reviews with them.

Where did the risks I have worked upon so far come from?

- I found them by researching the project documents and any others that may pertain to the context of the project; regional development plans, possible forms of contracts, case histories of similar schemes, earlier feasibility studies, records about previous uses of the site.
- Through discussions with the project feasibility people; the advisors, the experts, the promoters, the stakeholders and whomever else is around that may possibly have an insight I need to understand; the defeated generals, the opponents of the project, those who are weary of it all, those who may eventually have to operate, maintain or use whatever it is the project will deliver, etc.

I also visit the sites where the work will be done and ask myself questions about how whatever is to be built could be in this place, and what could go wrong. I freely take my own opinions and experience, somehow pulling from this investigative fossicking the set of risks I think ought to be modelled.

The situation is somewhat unstructured, I grant you. I have established a set of risks I have used to inform the funders and the PM, and yet here I am on the verge of standing up before a workshop full of design and construction experts ostensibly to identify a set of risks. The only way I can reconcile this is to recognise that such a workshop is both a formative exercise for them, and a confirmatory one for me. By the end of it, I will know either that my own set of risks was correct, or that it was not. In which case I will know what risks have been missed from the analysis thus far. The workshop attendees will have had the satisfaction of having contributed to the process, something that if denied would probably irritate them and return to haunt me. If I did have the set of risks right, they would gain an assurance that the risk is accounted for. Thus, though the event may be described as a risk identification workshop, in reality it is more of a review, and an important one.

Let me not skip over the matter of what to do when risks that have not been identified in any preceding stage surface at this one. The primal fear is that they do not feature in the analysis that has been used thus far to inform the funder and help the PM. The first thing to do is to let the newly identified risks stand, not ignore them in the hope they will go away. To do this would be to reject the professional advice of the workshop attendees, which is something the client has paid to have. The new risks should appear in the risk register and in the risk model so that the attendees can see that their advice has been taken into account. This mitigates any chance of 'we-told-you-so' happening later on in the project.

The next step is to think about the dilemma that has arisen. The new risks have to be modelled but the results of the modelling have already been published. Adding more risks would change those results. I am usually fortunate because having completed far more risk identification work than one review, and having taken both cost and time risks into account as well as estimating tolerances and transactional risks, I almost always find I have something in the analysis that covers the new item. Therefore when it comes to describing the modelling of a new risk, I can often state the risk is modelled elsewhere and explain how. The 'elsewhere' risk may have completely different description but I may be able to make an honest and transparent argument that it covers the new risk within its modelling. Thus:

> There is a shortage of C++ software engineers at the present time and premium rates may have to be paid to attract them to the project.

This may be identified for the first time during a review with the contractors but is arguably covered by a previously identified risk:

> There is an estimating tolerance of the software labour costs of 0 per cent to +25 per cent.

Or

> The ground in the cutting is waterlogged and may need draining and stabilising (new).

This is arguably covered by..

> There is a risk of landslips in the railway cutting (previous).

Indeed I would have difficulty defending myself against charges of double-counting if I modelled both the new and previous of the above.

On the (thankfully) rare occasions when I have missed a risk, I declare it to my client, though always in confidence. This news is theirs to hear first, no one else should know it without their permission. It is important to remember that it is their prerogative to play any political cards that are dealt them, even those caused by my error.

When I do miss risks, they are always transactional and almost always in late emerging documents I have not seen, particularly those that external agencies have been able to influence; letters of agreement with disaffected parties; the details of changes accepted to close contractual negotiations; undertakings given in consultation exercises. Missing these when they are significant is embarrassing but is part of the job. It is important to remember that without risk analysis, the funding solution would be even more unreliable.

In the next sections I will explain an approach to three different workshops:

1. one for the project engineers
2. one for the project managers
3. one for the project executive.

Workshop 1: The Engineers

I am no enthusiast for standing in front of an audience and I know audiences would rather have someone else who can entertain and interest them, but one must not cancel a workshop nor delegate its facilitation to someone more personable. Over the years I have developed a method which I hope will at least be thought thorough by the attendees.

I will begin by making an assumption that we are analysing a single project intended to deliver a single asset. The first question to research, which I usually answer myself after some background reading and thinking, is what professional engineering disciplines are needed to design and build the asset project? If it were a railway project these would be:

1. Civil engineering for the structures.
2. Railway track engineering for the railway lines.
3. Traction power electrification engineering for the overhead electrics.
4. Signalling engineering for the signalling.
5. Telephone engineering for the communications.
6. Systems engineering for the IT.
7. Electrical engineering for the lighting and low power systems.
8. Project management

I draw this on a board in the form of a breakdown structure (Figure 6.13). Other projects will possibly make use of other engineering disciplines and so for them I may need to exclude some of the above and to include others; geotechnical engineering, hydraulic engineering, chemical engineering, mechanical engineering and so on.

As you can see, I am looking for the sets of technical expertise that are essential for the design and construction of whatever asset it is that the project has to deliver. My criterion for inclusion of a particular expertise in the set is that the asset could not be designed nor built without it. I am aware I am excluding many important functions, but they are not essential by this criterion. For the time being I am assuming that the other areas of expertise are competent and sufficient to support the project, and that in consequence they do not present a risk to it.

It does not matter much if I am missing one key expertise from the list because this will soon be pointed out to me during the workshop. If this is the case I may have to supplement the workshop with a follow-up discussion with the expert I neglected to invite.

I think about the above in advance of holding the workshop, making inquiries into who is the best engineer in each discipline so that I can make arrangements for them to be there with their team if needed.

In the first stage of the workshop, I ask each expertise in turn to describe for the benefit of everyone else the items that will need to be constructed, built or manufactured in their domain. I call these the construction work packages. I also ask that they include any items that will be bought off the shelf instead of being made. I note them all and list them on the breakdown structure appropriately (see Figure 6.14).

Once I have this list, I then ask which of the items have to be designed. These are the design work packages. Within design, I allow anything that commences with a blank

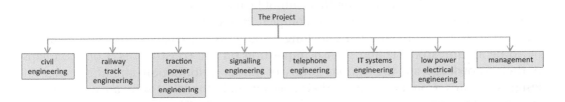

Figure 6.13 The first two levels

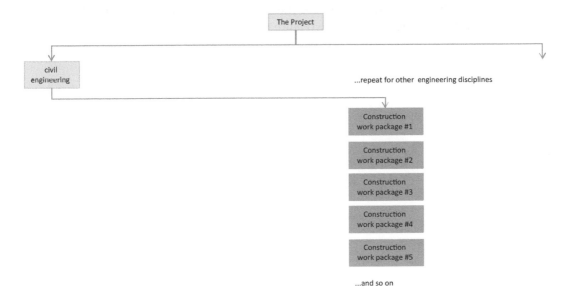

Figure 6.14 The civil engineering packages that need to be constructed are added

sheet of paper, to things that will be bought off the shelf. I classify the specification of these as a design activity, even if it is nothing more complicated than an instruction to purchase Brick A instead of Brick B. It is still a designer's decision. By this criterion, everything that is to be built has to be designed.

I then cross-link the design packages to the construction packages. This can be a one-to-many relationship, for example, one design for a bridge span may be intended for construction in several different places.

Once all the design packages have been identified, I take each discipline in turn and ask everyone attending to identify any design work packages that are superfluous, or any more that may be needed. From the analyst's point of view, which design package belongs to which discipline does not matter at this stage as long as they are all identified. Having a plenary session in which signalling engineers (for example) can ask who is designing the signal bases is important in ensuring all of the works are identified (Figure 6.15).

In Figure 6.16, design package 1 is used for construction package 1; 2 for 2, 3 and 4; and 3 for 5.

Next I ask the experts: do they know everything they need to know in order to commence the design packages? And do they have everything they need for the design and subsequent construction? I raise this last point with the designers in an attempt to prevent them designing something that cannot be built within the funds and timescales likely to be available. Once a design is complete it is very difficult for it to be undone. If it is then found to require twice as long to build as is available then that seems to me to be an unreasonable and foolish imposition on the constructors. Common gaps in their knowledge are the various forms of survey information and the choice of design standards. Common gaps in their needs include the latest software applications, shortages of resources and lack of manufacturing capacity. The provision of these I call the facilitation work packages because with them design and construction can go ahead, but without them that may not be possible.

I add the facilitation work packages to the diagram and cross-link them to the design and construction packages they enable (Figure 6.17).

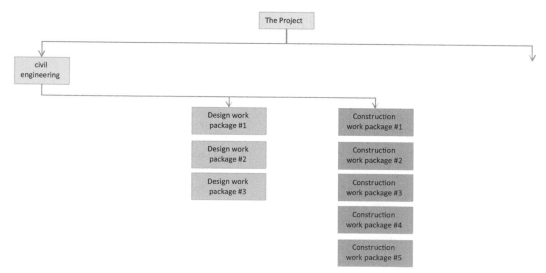

Figure 6.15 The design work packages are added

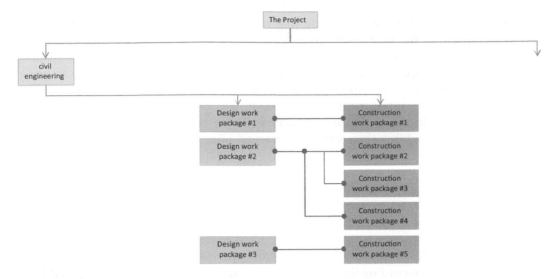

Figure 6.16 Design and construction work packages cross-linked

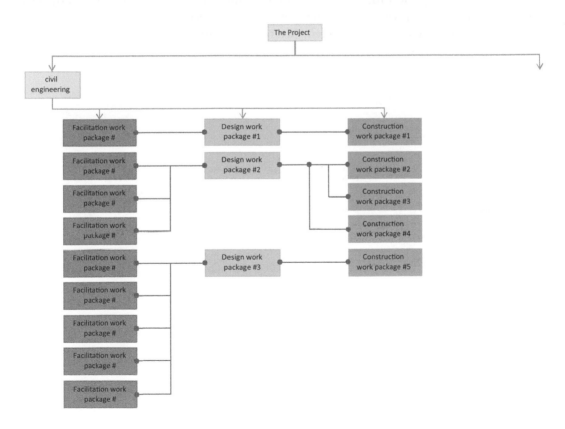

Figure 6.17 Facilitation work packages added and cross-linked

I now have a picture of everything that needs to be designed and built. However, not everything that has been built can be used straight away. Perhaps operators have to be trained, or perhaps a trial running period will be necessary to test and commission systems. In which case I ask, and list, the work packages that need to be done to bring what has been built into use. These I call the 'Into Service Packages' (Figure 6.18).

Notice how the interconnections between the construction and into service packages imply a degree of systems integration. Some into services follow on directly from construction, while others can only proceed once other into services have been completed.

I always research these four tranches for each engineering expertise in turn in an open forum. It does make for a long day, so I offer breaks after each discipline both to give the attendees a rest and to let me transcribe the breakdown structure to my notebook so that I can clean the board for the next session. If the attendees have got the idea – drawing things up on a board seems to help assimilate ideas more quickly than projecting them ever does. Perhaps this is a learnt behaviour from the way we were taught at school – I sometimes switch to recording the breakdown structure straight into an Excel spreadsheet that is projected onto a screen at the front of the room.

After completing the structure for the engineering disciplines I finally turn to the last expertise: identification of the management work packages needed to ensure the work packages can be enacted and delivered. My responsibility is to the funder, and if the

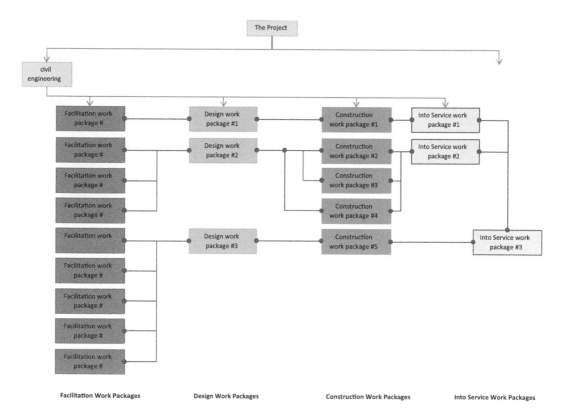

Figure 6.18 Introduction into service work packages added and cross-linked

company chosen to deliver the project is flawed in a way that could prejudice its ability to do this then I need to know. I do not want to take the risk of non-project managers silencing those who think there are problems that should be discussed, problems the funders may well be able to solve on behalf of the project. This is the reason why I do not want management (other than project managers) attending the review. Here are some examples of issues that would probably not have been aired had the reviews in which they were identified not been conducted exclusively among project staff:

Significant numbers of key experts are due to retire within the project timescales.

Manufacturing capacity is reducing through machinery degradation.

Inspection and review procedures are too time-demanding to accommodate within planned activity durations.

The design produced by the architects is too intricate to be built within the budgets and timescales available. It needs to be changed.

The assumption that all permissions required will be granted first time is contrary to experience.

The contract to be imposed on the constructors is too onerous. The stick is so big that the first time something major goes wrong for the contractor, it may as well give up trying.

We are paying over the odds for centrally supplied materials we could source locally for less.

You may pick up a hint of anger and frustration in their phrasing. The analyst needs to be sensitive to this, recognising that admissions like these can be career-limiting. I always therefore always endeavour to make sure the circumstances of their disclosure are confidential. I am always ready, like journalists, to protect my sources. Of course, I do not necessarily believe everything I am told without corroborating evidence.

In case you are interested, the above were all eventually treated as risks and were modelled respectively as a loss of productivity, a spread duration, an unresolved option, a rework cycle, a not-modelled but declared (obviously) and a spread on rates.

The resolutions of these problems are the management work packages needed to ensure delivery. If they had been resolved then they would have not needed modelling. Figure 6.19 shows the completed picture.

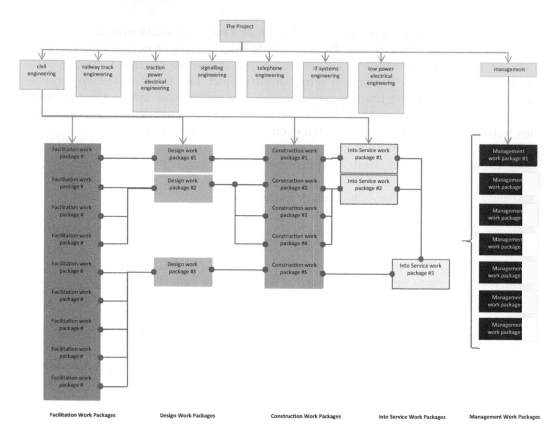

Figure 6.19 Management work packages added

I now have a complete work package breakdown structure of everything that needs to be done to carry out the project.

I then take each work package in turn and ask the attendees two open questions about it:

1. Can the job be done?
2. What could go wrong?

The questions are succinct so as to make it easy to recall them. Though they are in a semi-closed form,[2] they are not intended to elicit simple yes or no answers but rather to lead on to a wider discussion. Before answering them though, at this point in this imagined review, practicalities must sometimes be taken into consideration – the business of researching the work structure and, now, of moving on to discuss each package in turn, can hint at a long day ahead for the attendees. So having posed my two questions, I usually offer to end the review, provided it can be reconvened at a later date. This gives everyone a chance

---

2   The first question is closed because it tends to elicit either a yes or a no answer. The second question is open but it can elicit the answer 'nothing', and so I describe it as semi-closed. Both questions rely on the conversational ability of the analyst to generate fuller replies from the possible terse responses.

to think of the answers and also to go home early, which is always appreciated. Usually the subsequent reconvening takes the form of me going to see everyone individually to note their answers, a method I personally prefer. This way, the inquiry tends to end when the debate has been completed and not before. Others that are present, but who are not engaged in the discussion, start to fidget, and subconsciously will the discourse to move on. But before I end the review for the day, I always ask the attendees if they will think about the two questions in respect of all of the work packages and not just the ones in their area of expertise. This is because it is often the outsider who knows what is wrong with the insider's work. It is often our own the flaws that we never quite see.

As for the answers to the questions, any negative answers to the first question – can the job be done? – are usually potential threats to the success of the project and hence are its risks. Typically the negative answers reflect an inability or a lack of capacity of the project delivery organisation. Some common examples are:

- The procurement department cannot deal with the anticipated quantity of contracts to be tendered and awarded within in the planned timescales,
- The numbers of expert staff required cannot be found or employed or accommodated,
- There is no technical solution available,
- The work package contravenes the law or the applicable engineering standards,
- The organisation does not have people available to it with the skills needed.

It is easy to see how these and those like them can be written up as risks, and with a little more research into their behaviour, their modelling primitives chosen.

The answers to question 2 – what could go wrong? – tend to be event-type risks. For example:

- Because the ground is alluvial drift and gravel, the signalling bases may settle, tilting the signals and so putting the train driver signal sighting distances out of specification.
- No tenders may be received for the estimated cost because an important item was incorrectly assumed to not be required when it was prepared. Some examples:
  - A stock of maintenance spares.
  - The contractor will not have to conduct certain surveys because the employer will.
  - Materials will be supplied from the employer's stockpile.
  - Land will be free-issue, i.e. paid for and supplied to the project and its contractors by the project promoter.
- The proposed works are unpopular and are likely to disrupted by objectors.
- The embankment has a history of landslips and some may occur during the project time frame,and so on.

If answers to the two questions have emerged during the review I write them up immediately. Once I have a good set of notes, I am thankfully no longer dependent on the vagaries and contradictions of bullet points on flip charts for the drafting of the risk register and the encoding of the risk model.

Workshop 2: The Managers

I have found over the years that the engineering community tend to see risks where others may not. The others are sometimes not wrong. For example, the risk given above that the numbers of expert staff required cannot be found or employed or accommodated.

This may be a very real problem to a forward-looking engineering manager, but not one at all to a commercial manager who intends to contract out the work to an overseas supply chain. Therefore I need to cross-check the results of the first workshop in a second workshop attended this time by the managers. Depending on the risks identified I may invite experts from procurement, commercial management, operations, project services, human resources, accounts, finance and IT services, to confirm or deny the existence of the first workshop risks. They are usually busy people with little time to spare, so I usually let them know that risks pertaining to their specialised field of expertise have been identified in an earlier workshop.

In Workshop 2 I give each attendee a list of the Workshop 1 risks, and my explanations of how I propose to model them. I then lead the attendees through each risk in turn, noting their comments. Where there is a clash of opinion between the engineers and the managers, the doomsayers and the heroes, it is for me, the analyst, to weigh the evidence and to decide which view or which compromise the analysis should reflect. Analysts must remember that their objective is to get the right answer and not the warm appreciation of those attending.

Workshop 3: The Executive

Although this is a workshop that may be difficult to arrange it can bring substantial benefit, because it may give a seal of approval to the outputs of the previous two workshops; it may result in executive assistance being forthcoming for the prevention of risk; and it may head-off the threat of an executive that has not been informed about the risks. When this happens the analyst sometimes only finds out when asked for an update on the management of a risk they knew nothing about.

Most importantly of all, the workshop will be attended by those who generally have a great depth of business experience, and intimate knowledge of the corporate development strategy and goals. The attendees may therefore identify further risks, but equally they may also be able to explain why other risks of great concern to the engineering and management reviewers should be closed.

These meetings may be difficult to arrange, but I have never found it so. Executives tend to be schooled both in the importance of corporate governance and the numerous codes associated with this, and so as a matter of proper, professional practice they are generally keen to find out what the risks are. Furthermore they are used to working as a multidisciplinary unit, unlike the engineers and managers who tend to work in silos surrounded by co-practitioners. Usefully too, executives tend to have a secretarial support network that is proficient in arranging their disparate diaries so that meetings can be held. The form of the workshop is simple – a guided walk and talk through the analysis and modelling of risks identified in the previous workshops, and in spite of many initial trepidations, I have always found executive risk workshops the easiest to set up.

There is a tendency for the process of risk identification to be described as a continuing one: that the register is a snapshot in time, that the process is cyclic (identify – analyse – manage – identify, and so on). This ongoing situationality serves the interests of the analytical fraternity, of whom I am one, more than it does the users of risk analysis: the funders and managers. It is to a degree both an admission that the work of identification is not complete and that its validity is in doubt. I have often heard myself speak from the back foot to excuse my work when it has been challenged, saying it is but a snapshot in time and its development is continuing. This I suggest was weak because no one present gained any assurance from hearing it, and I suggest that in order to avoid being placed in a similar position analysts should try hard to finish the identification process as soon as possible.

There may be detailed specificities that could be usefully added to a broad brush description such as 'landslips may happen' or 'there may be penalties for late delivery'. But to declare existence of these for the first time after an initial round of identification, analysis, modelling and management has been enacted is negligent to a degree. I know risks will be missed, but the excuse for this should simply be: 'Sorry but I missed them', not 'I have not got around to them yet'. No quantity surveyor would expect to forget the concrete and likewise no analyst should neglect the estimating tolerance on its quantity. As much as analysts need to get on with their job, they need also to get it finished without hiding behind a continuing process.

I mention risk identification at this point in the process because having notionally presented the risks in a workshop to an executive team, it may happily accept a second review to confirm their input has been incorporated. But my experience is that requesting a third meeting to identify any further risks will be seen as a third strike with the obvious connotations.

This need to identify all the risks is the reason why old hand analysts given a short study tend never to plunge straight into model and database design. They start to read and draft lists of risks, they start to make calls, to have short discussions with experts and to write up more risks. They know they are more likely to be given sufficient time to model their findings if at the end of their commission they can show a complete and thorough description of the risks.

And so I always try to bring the workshop process to a close by publishing my findings. I notify the attendees that the risks they have identified have been included in the risk analysis and I give references to where this information is recorded. The only modification to what I intend to be a generally transparent set of records is that some risks may be commercially sensitive, and so I may design a risk data structure that has restricted access.

7 *A Risk Analysis Process*

The world of work is populated not only by people paid to be doing it, but also people who are paid to be concerned in some way about the doing of doing it. They write specifications, procedures and standards, and design process maps and forms that others have to complete (why do they never fill them in themselves?). I see this as a Maslovian[1] need to establish a role within the crowded domain of project management, wherein there are many more people than projects and so not everyone can be a project manager. They are motivated by economic self-interest to find a job that is useful, hopefully valuable, that will give them a secure and sustained position within the pantheon of project management services. Being the risk manager is one of these roles, but it is one that contains within it two distinct remits, one of which is allowed to grow while the other is quietly let go. The remit that grows is to design, establish, roll out and operate a process of risk management on a project or for a company delivering many projects. This is a good thing but, from the point of view of the funders, it is not quite the best thing. The other remit, the one that is to a noticeable degree not done, is actually managing the risk in the obvious sense of taking direct action to prevent, contain and remedy it. An argument is made that risk managers are not best placed to do this, or that they are not empowered in some way. Instead of challenging these limitations and overcoming them, too often on projects, risk managers simply record risk.

If a manager were acting to reduce the risk, they would be doing calculations to help choose between reduction actions competing for limited resources, and drafting clauses for contracts that would ensure prevention actions are given the force of contractual obligation. They would be ensuring the people, machines and materials that may be needed to contain the unwanted outcomes are available, and they would be possibly controlling the budgets that may be needed to pay for remediation. Someone carrying out this kind of risk management – taking action to reduce it – is taking a personal risk that on any day they may be found to have got it completely wrong: prevention was not assured, the wherewithal needed to deal with an emerging problem was not in place, and there are no funds readily available for remedy. Such responsibilities are real, and failing them would probably be conducive to an appraisal of the manager's suitability for the job.

It may be more in one's economic self-interest to have a role that is less in the firing line, which is to be a risk manager in the sense of managing the process of managing risk through the design of risk procedures, line management of risk personnel and the establishment of IT systems for recording the risk. The worth of such work cannot be denied because it is necessary in all organisations managing projects, but it surprises me how many more people are engaged in managing the process of risk management than

---

1   A. H. Maslow (1943) A theory of human motivation. *Psychological Review*, 50, 370–396. See http://en.wikipedia.org/wiki/Maslow's_hierarchy_of_needs.

there are engaged in managing the risk itself. There is some intriguing indirect evidence of this. David Hillson[2] gives a presentation in which he shows process charts for risk management process after risk management process, all of which contain procedures for identification, analysis, management and reporting of risk (usually repeated cyclically). But none of these require the process operators themselves to do anything about it, i.e. to reduce it. It reminds me of the well-worn comment about consultants being people who tell you the time using your own wristwatch. I often sense that risk is what sustains some risk people – they ask you what it is and report it back later. Reducing it, or even better, eliminating it, could be a job-reducing strategy. An image of a doctor nurturing, rather than curing, ailments comes to mind.

The management of the process of (somebody else) managing the risk has grown to become a big battalion in the field of risk work. Google 'risk management standard' and 'risk management procedure' and you will find many to choose from, yet despite this, every large client I have had has at some point in its corporate development, instructed some of its staff to write one specific to its own needs. When scrutinised, this seems hardly different from any of the others, save that the reporting procedures will be customised to the organisation design. Dr Hillson found 15 standards for risk management and wrote a comparative review of some these and four more:[3]

Standards (after Hillson)

1. Association for Project Management (2004) *Project Risk Analysis & Management (PRAM) Guide*, 2nd edn. High Wycombe, APM Publishing, ISBN 1-903494-12-5.
2. Association for Project Management (2008) *Prioritising Project Risks*. High Wycombe, APM Publishing, ISBN 978-1-903494-27-1.
3. Association for Project Management (2008) *Interfacing Risk and Earned Value Management*. High Wycombe, APM Publishing, ISBN 978-1-903494-24-0.
4. British Standard BS 6079-3 (2000) *Project Management – Part 3: Guide to the Management of Business-related Project Risk*. London, British Standards Institute, ISBN 0-580-33122-9.
5. British Standard BS 31100 (2008) *Risk Management – Code of Practice*. London, British Standards Institute, ISBN 978-0-580-57434-4.
6. British Standard BS IEC 62198 (2001) *Project Risk Management – Application Guidelines*. London, British Standards Institute, ISBN 0-580-390195.
7. Institution of Civil Engineers, Faculty of Actuaries and Institute of Actuaries (2005) *Risk Analysis and Management for Projects (RAMP)*, 2nd edn. London, Thomas Telford, ISBN 0-7277-3390-7.
8. Institute of Risk Management (IRM), the Public Risk Management Association (ALARM), and Association of Insurance and Risk Managers (AIRMIC) (2002) *A Risk Management Standard*. London, IRM/ALARM/AIRMIC.
9. Institute of Risk Management (IRM), the Public Risk Management Association (ALARM), and Association of Insurance and Risk Managers (AIRMIC) (2010) *A*

---

2   http://www.risk-doctor.com/

3   *Risk Management: An International Journal* 2005, 7 (4), 53–66 Dr Hillson's web site has recommended reading lists. I am grateful to Dr Hillson for the providing this list.

Structured Approach to Enterprise Risk Management (ERM) and the Requirements of ISO 31000. London, IRM/ALARM/AIRMIC.

10. International Organization for Standardization ISO 31000 (2009) *Risk Management – Principles and Guidelines*. Geneva, International Organization for Standardization.

11. International Organization for Standardization Guide 73 (2009) *Risk Management – Vocabulary*. Geneva, International Organization for Standardization.

12. National Standard of Canada CAN/CSA-Q850-97 *Risk Management: Guideline for Decision-makers*. Ontario, Canadian Standards Association, ISSN 0317-5669.

13. Project Management Institute (2008) *A Guide to the Project Management Body of Knowledge* (PMBOK®), 4th edn. Newtown Square, PA, USA: Project Management Institute.

14. Project Management Institute (2009) *The Practice Standard for Project Risk Management*. Newtown Square, PA, USA: Project Management Institute.

15. UK Office of Government Commerce (2010) *Management of Risk: Guidance for Practitioners*, 3rd edn. London, The Stationery Office, ISBN 978-0-11-331274-0.

16. IEEE Standard 1540 (2001) *Institute of Electrical and Electronic Engineers, American National Standard*.

17. CEI/IEC 62198 (2001) *International Standard Project Risk Management Application Guidelines*, 1st edn. International Electrotechnical Commission.

18. JIS Q2001:2001(E) (2001) *Guidelines for Development and Implementation of Risk Management System*. Japanese Standards Association.

19. AS/NZS 4360 (2004) *Risk Management Standards Australia/ Standards New Zealand*.

This is intuitively a lot of effort expended, not only on process development but also in the printing and publishing of the outputs, rolling them out, training, monitoring and, dare one say it, promoting them against their rivals. When say the tenth risk-management process had been published, why did someone then decide that the business of risk analysis and management would be advanced by writing an eleventh? The further development of the risk-management process is really only justified to the extent that any additions and adjustments incorporated will support those risk analysts and risk managers who may be proved wrong in their advice and actions.

The development of the risk-management process has developed a new arm in recent years: the introduction of enterprise-wide risk management IT systems. I am sceptical that investment in risk management IT instead of in people who act to reduce risk produces the benefits project funders seek. No matter how attractive the systems may be to IT departments and managers, such systems do not themselves deliver the insight into a funding requirement or the solution-led risk management that a risk analyst or risk manager can. The systems can help organisations provide evidence of a responsible attitude to risk and help their staff keep proper records, but risks are dynamic. Their exposures change with better understanding and vary over time. The discrete exposures of a group of risks may overlap, so that the funding is double-counted, or it may contain gaps, such as the funding is missing. Human intelligence is needed to make sense of it all. To replace risk analysts and managers with risk management IT systems is a mistake because, though an enterprise wide risk system may be able to tell a client *everything*, it is the *something* that a client really wants to know, and this requires an analyst to spot it, explain its significance and then act.

To make the need for a process of risk management a justification for the safe and sustained existence of a cadre of staff developing process and recording risk, however sophisticated these may be, with a possible investment in risk management IT systems is not an assured way to reduce risk. It is the person who acts that makes the difference, not an IT system or a process.

Having argued that the remit of the risk analyst and the risk manager should be to do the job I think clients want done: reducing the risk, I want to correct any impression I may have given that risk analysts and risk managers can therefore ignore process and proceed about their business in an ill-disciplined and mercurial way. A risk analyst or risk manager has only one goal: to deliver the right results, and to achieve this they need to avoid situations in which they are obliged to excuse their work on the grounds that they did not know something that was critical to it. This talisman: 'no excuse for not knowing', necessitates a disciplined, thorough and complete way of going about the business of risk analysis, so it may be useful if I close this book by setting out the approach I use to provide results I believe the project funder and the project manager will value.

Risk Workshop

Traditionally, risk analysis begins with a risk review in which the analyst stands before an assembly of those who have time to attend, flip chart to hand. The analyst then tries to lead and sustain a discussion on the identification, description and quantification of the risks to the project in question. Many find this a difficult thing to do, dreading the emergence of a moment when they begin to realise the attendees are bored, or that their own lack of knowledge about some aspects of the project is being displayed to manifest irritation.

Bad performances in front of the flip chart are not the biggest threat to the success of a workshop. More important are flaws in the design of the workshop, and in the mental preparedness of the analyst, both of which it would be prudent to pre-empt. In what follows I will set out some arguments to bear in mind as well as some practical suggestions that should help. The most important thing is that the objective of a workshop is not to win over the attendees but to get information from them. All a risk analyst has to do, philosophically speaking, is provide the right answers. Get this right, and analysts should find that a modest ability in front of the flip chart will do.

First – while the workshop attendees may be a forum, it may not be one of experts, and even if it is, the experts may be wrong for the subject of the analytical inquiry about to be conducted. It is therefore critical to find out (in advance) who has the information needed, and make every effort to have them attend. This is usually done by choosing a date that suits them and, I suggest, by asking them personally. No one likes to be summoned to attend a workshop solely by a broadcast e-mail. Making a personal approach that includes a short discussion about the workshop: what it is for, who else has been invited, asking what things could be usefully covered in an introduction to the day (a tutorial on time risk modelling perhaps), will mitigate the tensions that tend to arise just before and after the start of the workshop. In these circumstances, the analyst will be working with people even if only to a limited degree, that they are aware of.

Secondly – there may be a power hierarchy hidden within the attendees that inhibits quieter voices speaking or messages that are anything less than positive being uttered.

This is especially true when a workshop discusses risk quantifications, when possible exposure to failures of management will sometimes be denied by strong voices in the face of historical evidence to the contrary. Some common examples are that procurement processes are frequently bottlenecks, design approvals tend to take longer than planned and works do tend to take longer to complete than intended.

Despite this, strong voices will sometimes assert that it will be different this time. I doubt it. A bias towards an unsubstantiated optimistic view of risk should be pre-empted in a welcoming introduction by saying that risk is in the project, not the people, and especially not in the attendees. This is a moot point but also a pragmatic one to make when trying to get people to discuss risk openly. No project is risk-free, but if strong voices persist in holding a view that their part of it is risk-free, that it will be delivered on time and to budget and so forth, then the analyst should consider openly declaring a difference of opinion, an intention to communicate this to the funders and a resolve to model the apparent optimism.

If on the other hand the strong voices felt on consideration that some funding against things that may go wrong would be prudent, then perhaps the workshop will be able to discuss the what, why and when so the funders' can make an appropriate provision. After all, the money the attendees may need to spend on risk will not be theirs, and the parties who do own it may be perfectly willing to have it spent in the right circumstances.

On the moot point of risk not being the people but the project, assertions that this is not so and that someone is the cause of a risk, should be noted for further research later, ideally on a one-to-one basis if it is sensed that further investigation of the allegation would prejudice a successful workshop.

On the matter of strong voices denying the existence of risk, thereby subliminally forcing quieter voices to cede to their opinion, analysts should remember there is no excuse for not knowing. Try to arrange to interview them at another time in case the insights and experiences they have are be important.

Thirdly – this is possibly devastating for an analyst standing up in front of a flip chart – the voice that questions the process. This decrying is usually done in one of two ways: either the ability of the analyst to understand the subjects being debated is doubted, or the validity of the result (the cost curve, say) is called into question before it has been computed. The challenging comment of 'rubbish in, rubbish out' never prompts a sharp retort along the lines of 'then do not provide it' because the validity of data is an ever-present concern of risk analysts and attendees. Some observations may help.

- It is difficult to discern good data from bad during the initial gathering of it, but bad data is often easier to identify once a risk model has been fully populated and results produced. Some quantifications may be seen to be double counts while others are seen to be missing.
- Results often attract the attention of people senior to workshop attendees who are aware of things that cancel or reduce risks, or the opposite.
- To describe data as rubbish because it is an approximation at the time is wrong if the variability exists. To use a precise quantification when one does not exist because of a preference for a spot value would in fact be to put 'rubbish in'.

Perhaps the most effective counter to the rubbish in, rubbish out challenge is the observation that good risk modelling justifies typically a third of the funding, and a quarter of the time a project needs, before posing the persuasive question: would the project team like to proceed without them?

This in turn may provoke the question: why should the one third and the one quarter not be added to the estimate and the plans, and the workshop closed? If this was standard practice then the contracting community would soon become aware of it and increase the price of their tenders over time to absorb the additional funding. Consequently, funds intended for risk management would instead contribute to higher profits, higher overheads and more relaxed management regimes, and yet the risks would still be there, though now presumably with their management unfunded.

Fourthly – flip chart notes have a half-life of roughly an afternoon. After that time, half the comments and sentences without subjects will have lost their meaning and importance. This is a no-way-out situation that analysts should avoid by announcing in their introduction – though they may write subject headings on the flip chart to remind everyone what is being discussed, they will be spending a lot of time sitting down taking notes. No analyst should take the risk of concluding a workshop to find they have an inadequate and incomplete set of notes. I personally never take the chance of relying on someone else's notes or minute taking, and so I apologise in advance for my frequent withdrawals from the discussion to make my own notes. Some colleagues have tried recording reviews but have found the one thing not picked up by the machines is their own thinking. They do think it is a good aide-memoire though, one that over time improves in comparison to their notes.

Fifthly – workshops tend to be long sessions and it is impossible to sustain the attention of everyone all the time. Attendees with perceptive points miss the appropriate moment to make them; shallow comments are made; subtle pertinent nuances and caveats are condensed into monosyllabic responses; information is withheld to avoid prolonging a line of inquiry, and peoples' attentions wander. A common approach to avoiding such scenarios is to break the workshop up into work groups, each tasked to identify risks. Might I suggest that this is the wrong approach? Working group sessions always conclude with a plenary session at which often little or no attempt is made to criticise or challenge contributions no matter how much they warrant it. There is a tendency to vagueness rather than being enlightening. I therefore suggest that analysts should avoid breaking a workshop up into work groups because the risk to their own understanding of the risk is too great.

A good way to re-stimulate a workshop is to change the subject being discussed to one that everybody attending will have an opinion on. Questions along the lines of 'what should X do then?', or 'how can Y help you avoid this?' or 'what should be done to constrain Z?' not only revive a flagging meeting but are productive in themselves and conducive to revisiting more difficult questions, those along the lines of 'why do you think nothing will go wrong for you?'

Sixth – drawing from my own experience, I have never known a long, dreary, workshop lead to an affirming and appreciated set of results. Whenever I have lost the support of the attendees in the workshop, I have never won them back when I have presented the risk register and the cost curves. And yet their support is vital. The second thing funders want to know after seeing the outputs is do the workshop attendees agree with them? Consequently, I put more thought and effort into gaining the agreement of the attendees than all other aspects of the workshop. Tried and tested over many years, I have arrived at a method that works for me and that I always try to arrange to have happen.

If, as often happens, a client is only prepared to grant me a single workshop and not the three described earlier, then it seems to me wasteful to risk it on a risk identification and quantification workshop that may not go well.

What I prefer to do is use my one big bang workshop to get acceptance of the results, and not to identify the information that goes into them. I may then be able to give to the funders an assurance that every department or team connected to the project has seen the results and (hopefully) agrees with them. This is a valuable output from the workshop, much more so than a bundle of flip charts and a disgruntled audience, neither of which is of any value to the funders. This approach puts the risk workshop at the end (WAE) of the study, not at the start start.

WAE means the risk identification researches and risk modelling have to be done before the workshop takes place, but this in turn means the research will be done more likely than not on a one-to-one basis, with proper time and attention given to each interviewee. A 1:1 approach gives an analyst the opportunity to concentrate on each interviewee's specialist area and an opportunity to show they know something about their own, something workshop at the start (WAS) generally denies. Interviewees are given a chance to speak without the influence of others and to produce evidence such as drawings, reports, site visits, equipment, photographs and so on that generally do not appear with WAS. They will have been given the opportunity to contribute fully to the analysis, and the analyst's understanding of them and their work will have improved in consequence. I have found WAE leads to a better risk analysis and has even created an atmosphere of mutual trust and cooperation that dispels any perception on the part of the would-be attendees that risk analysis is waste of time. Analysts who use WAE should find themselves standing up in front of people who know them and who are interested in their work, rather than those who doubt it.

The single major improvement WAE brings over WAS is to the validity of the analysis. By letting attendees see the inputs and the resulting outcomes at the workshop, my experience is that corrections and adjustments are voiced on the spot, and are done so in an atmosphere that is positive and supportive. This is even more the case if the risk models can be run live in real time so that the impact of suggested changes can be seen quickly. It may sound perverse at first, but if an analysis can be seen to be wrong then it is more likely to be right, in the sense that the implicit clarity with which it has been done will be conducive to its correction and imminent validity. A WAE analysis may well not be correct on its first presentation, but given the inherently supportive scenario of WAE, it soon will be and this will provide the benefit the client it wants: the agreement that the cost and time curves, the ownership analysis and the risk management solution are supported by the workshop attendees.

Finally, whether workshops are WAE or WAS, an analyst must always remember that they represent the interests of the client, and that of those attending a workshop, they may be the only one with that responsibility. If the analyst does not speak up, or doggedly persist with a line of inquiry, no one else involved may do so.

Part 1: A Risk Process to Inform the Funders of Their Risk Exposure

The process is a set of sequential steps for WAE and not the more commonplace identify – analyse –action – repeat cycle, which I eschew in the belief that it is important for the credibility of risk analysts that risk analysis should have a clear start, deliver something of worth, and then finish. It is up to the client whether or not to repeat the process and analysts should only be given further work because the client wants it done, not because they have managed to attach themselves to a cyclic process enshrined in a company manual that cannot be gainsaid no matter its value.

1. Hold a preliminary meeting with the client to discuss what the risk exposure curve sought is for. Is it all of the project or a part of it? Is it the capital costs alone or are operating costs and revenues to be included (these I will discuss in a subsequent book)? Is a safety risk assessment to be included or is this being done by a specialist department? In short, scope the study.
2. Obtain lists of the relevant documents and the permissions to read them. Include those documents that are from outside the project record but potentially relevant: local authority policies, lessons learnt reports from projects that are similar in some regard and so on.
3. Obtain lists of experts and stakeholders to the project and their contact details.
4. Ask the client to notify document holders, experts and stakeholders about the study and to give them approval to talk to you and hand over documents.
5. Prepare a schedule of meetings and ask the client to arrange them.
6. These arrangements will take time so in the meanwhile, prepare first drafts of the report, and model any presentational material. These may contain entirely fictional narrative, empty tables and false figures but I find the time spent preparing them pays dividends later when deadlines approach. In particular, I set up and test output data routines that will automatically produce the figures I think I will need with their correct labels and with my preferred layouts, sizes and colourings. I also plan the structure of my reports and generate their tables of contents, and then sense-check them. I often write an entirely fictional but realistic executive summary not only because it helps me design an analysis to fit the story I imagine I will be telling, but also because modifying it will be far quicker than having to write it from scratch when the study deadline looms. The quality of the written English will be better too because a period for the narrative to mature, to have lost its rough edges and non sequiturs, will have passed.
7. Go and get the documents. Read them and take notes. In particular, study the documents within the commercial frameworks of contracts and funding agreements for transactional risks.
8. Obtain copies of the estimate and the plan for use as the foundations of the risk model. If either of them does not exist then:
9. For the estimate, be prepared to compute the risk exposure function without an estimate and annotate presentations of its results with a caveat so that they do not include estimating tolerance risk.

10. Or the plan, research and produce one. If it cannot be done, be prepared to compute the risk exposure function without time risks and annotate presentations of its results with a caveat that they do not include the effect of time-dependent costs.

11. If there are work sites, visit them and look for potential problems. Take photographs of these. In particular think about: site access, existing services such as water, gas, sewage etc. And needed services, how the finished asset will be operated – and what problems could arise with them?

12. Prepare to interview each expert. Study their subject. Prepare lists of possible risks as you see them. Rehearse lines of inquiry.

13. Interview each expert and discuss the risks that they foresee and those that you foresee. For those risks that withstand scrutiny, discuss how they should be modelled and what their quantifications should be.

14. After each interview, build the findings into the risk model and risk register.

15. Continue to interview the experts, updating and re-running the model with each. Check that the additions to the model produce explicable changes in the results. Begin to get a feel for what the cost curve looks like and its important statistics. Note the reasons for unusual outcomes as they appear.

16. Begin to build up a story board presentation pack on research done and results calculated.

17. Prepare the workshop: print material, book projectors, arrange for whiteboards, buy marker pens, obtain a power distribution panel, tag evidence in documents, anticipate supplementary questions and prepare answers to them, prepare a confidentiality protocol for attendees to sign if commercially sensitive results are to be presented.

18. Rehearse the workshop: explain who is attending and why, present the results, explain why they are like they are, explain second-tier findings (rankings, ownerships etc.), seek confirmation.

19. Offer and give if wanted, a private, prior presentation to the client. Take into account their wishes in regards to disclosure of sensitive material etc.

20. Before the risk workshop takes place, visit the room and arrange it so that everyone can see what you will be presenting and can, to the greatest extent possible, see each other face to face so that discussion is helped not hindered. A lecture theatre is the worst possible arrangement, a round table the best.

21. Hold the workshop. Expect, encourage and note every correction given. If at all possible incorporate these into the model interactively to demonstrate their effect. Seek consensus on results. Give priority to note-taking. Be prepared to ask the attendees to reprise a discussion missed because of note-taking.

22. Close the workshop: give thank yous, recover confidential material from attendees, clear the room.

23. Before leaving, sit quietly and write up the notes. If at all possible, complete the register and model. Back-up the work before switching off the computer. Further if at all possible, complete the report the same evening.

24. The next day: meet the client and talk them through the report. Ask for a little time to tidy up and check workings.

Part 2: A Risk Process to Inform the Project Manager of a Viable Risk Management Plan

This process has two phases. In the first, steps 1 to 14, the analyst is seeking to inform the project manager which risks they are responsible for managing, and to ensure the funders know this and have funded the exposure. In the second phase, steps 15 to 20, the role of the analyst is moved slightly away from the numbers and more towards taking on management responsibility for assuring the actions in the risk management plant they have devised are carried out.

1. Complete the risk model and write a report on its results.
2. Analyse the ownership of each risk.
3. Amend the risk registers for ownership interface violations and funding blocks.
4. Report these amendments to the funders.
5. Partition the risk register according to the ownership analysis, and compute the risk exposure or cost curves for each project manager being supported.
6. Draw up a list of extraordinary risk management actions for prevention, containment and remediation for each manager.
7. Brief each manager on the ownership analysis, checking that contingency funding for the extraordinary actions is available to them. If it is not then seek a means to make it so. If it cannot be done then inform the funders of the situation.
8. If permitted, inform other stakeholders to the project of their exposures and confirm if possible that they have sufficient reserves to bear them. Confirm this to the project manager and to the funder. Explain all exceptions.
9. Check that all risk management actions are contracted. Confirm this to the project manager and funder. Explain all exceptions.
10. Rank the risks in each project manager's ownership by mean, correlation, urgency, trend and risk reduction leverage.
11. Assess the rankings and decide a rational priority for managing risks. Discuss, amend and agree this with the relevant project manager.
12. Confirm each project manager's completion date and budget.
13. Solve the risk model for these and decide what would be practicable risk-reduction action plans. Advise the funders if one cannot be found.
14. Discuss and agree with each project manager a risk-reduction action plan. This will obviate the need to manage risks according to their ranking.
15. Take the actions from the plan and record this in the project risk management data system (whichever form it takes).
16. Brief the actionees on the plan and the part they play.
17. Take note of any problems or difficulties anyone has with this and report it to the project manager.
18. After one to three months, review the plan with the actionees and confirm to the project manager that it remains viable and is underway.
19. Prepare applications for contingency fund withdrawals in advance in case they are urgently needed.
20. Be vigilant. The contingency fund is somebody else's money and if the analyst does not act to ensure it is used wisely it may not be.

Conclusion

I have tried to show that the amount of money needed to deliver a project cannot be computed by an estimate alone, nor can the amount of time that will be needed. The moment risks are identified in a project, the estimates of the costs and timescales required need to be re-calculated to take them into account. The calculations mature from being arithmetical in nature (the estimates) to mathematical (the risk models). This is a matter of necessary sophistication, though as I hope I have shown in sufficient detail, it is not a difficult one to master. Not doing the risk calculations will almost certainly result in the project being under-funded and under-timed and hence more likely to become overspent and late. The implication of this is that the amount of time and funding needed will both probably increase. However, I think project funders would prefer to have projects properly funded than to have them overspent and late.

Doing risk calculations for a project increases the funding required, but this does not necessarily result in the project costs increasing. The increased funds are for the risks, and risks are as realistic and describable as the labour and materials of the estimates. Not doing a risk calculation implies that costs of any risks whose unwanted outcome occurs have either to be met from the estimate or from additional funding. The overall outcome is the same: the funding has to cover the labour, materials and risks, and I suggest it is far better to make provision for the latter in advance of it possibly being needed. This will allow risk-controlling actions to be funded and contingency budgets for remediation to be hypothecated. This is a better strategy for efficient and effective management than to be driven by expediency and crisis management as an unfunded risk position would have to be.

Increased funding to cover risk should not result in a variation of C. Northcote Parkinson's Law, in which project outturn costs increase to meet the funding available, prevailing because the analysis justifying it can be used to devise an evidence-based solution for its management, that is to say its elimination or limitation.

The current largely management-led approach to risk on projects does not derive the insights: the exposure curve, the funding requirement, the probability of completion, the ownership and the risk-management solution that an analysis-led approach does. I hope this book will nudge the business of project risk towards becoming analysis-led in future. As my colleague Peter Harnett says when asked to comment on the validity of a management-led risk analysis: 'deliver it for amber then'. Analysis-led is much the better way, and reassuringly, it is much more difficult to do.

I suspect projects do not become overspent and late because the project managers did not do their job well enough, nor the planners, designers and cost managers theirs. They become overspent and late because the unwanted outcomes of risks occurred causing the budgets to be exceeded and the deadlines to be passed. Projects therefore overspend and overrun because risk analysts do not do their jobs well enough. There may be good reasons why. Perhaps they were not asked. Perhaps they were not given enough time. Perhaps they were not given access to the right people, places and information. But my hope is that by writing this book, it will not be because they did not know what to do.

Appendix 1
Risk Identification Checklists

There is no scheme, structure or logical sequence to the following risk identification lists. They are an ad hoc collection of lists I have stuck into an e-scrapbook over the years. They have come from my own thinking and from that of others whose names I have forgotten but to whom I owe an acknowledgement.

They are ungrammatical and overlapping and so I use them more as prompts to my own thoughts rather than as checklists of specific risks. I prefer this approach because it makes me think about the project I am going to analyse and because, if they were a checklist of specific risks then there is a danger my analysis would degrade into something prescriptive wherein the only risks that pertain are the ones on my list.

Before I start a workshop I quickly read through the following and highlight those I think may be productive lines of inquiry to bring up in conversation. I often repeat this during the first interval in a workshop when I have an idea of which areas the interviewees are keen to discuss, so that I myself am not unduly influenced by their preferences regarding lines of inquiry.

List 1

1. Estimating uncertainty in the choice, quantity and cost of people, machines and materials.
2. Unresolved options on design, construction method, choice of materials.
3. Scarce resources leading to cost escalation and the scheme then needing to pay premium costs to get done.
4. Works elsewhere caused by the scheme (knock–on).
5. Inadequate access.
6. Payments to licensees.
7. Compensations to third parties.
8. Payments to stakeholders for services.
9. Environmental mitigation e.g. protection of..., relocation of..., avoidance of...
10. Disruption.
11. Objections and rejections on safety grounds.
12. Conflict with other projects.
13. Dependence on other projects.
14. Adverse impact on other users of infrastructure we may we working on or about.
15. Shortage of plant, trains and expertise, of...

16. Bursts ... one small trip up and major costs or delays will ensue, e.g. the next time the asset is planned to be removed from operations for the project to access or use is three years away.
17. Costs and time needed for legal, public or political consultation processes.
18. Costs and time needed for negotiation or establishment of partnering processes with co-developers.
19. More, or modified, IT will be needed.
20. Incentives will be needed to get the scheme done.
21. Unusual costs associated with introducing the works into service e.g. coordinating replacement services will be needed, or rare spares will have to be stockpiled.
22. More, or changed, maintenance.
23. Unusual and difficult disposal problems (in due course).
24. Unusual costs associated with eventual renewal or replacement.
25. Incomplete knowledge on, say, condition, ownership or liabilities that might necessitate money and time to evaluate and resolve.
26. Prior repairs or restorations or reinstatements might be needed to facilitate the scheme.
27. Public and political consultation costs.
28. Legal costs: services, fees, commissions.
29. Financial charges: services, fees, commissions, interest charges.
30. Are there any contractual penalties?
31. Are there damages payable which are likely to be capped, leaving the client exposed to the excess?
32. *Force majeure.*
33. Inflation.
34. Taxation.

List 2

1. Who is the client/sponsor?
2. What do they want?
3. Can it be done?
4. When will it be done by?
5. What will it cost?
6. What will it cost if it goes wrong?

List 3

1. What can't be done?
2. What could go wrong?

List 4

1. As you see them, in your own words, what are the risks?
2. Describe five feasible situations which would lead to *over spends*:
3. Describe five feasible situations which would cause the project to *go late*:

List 5

1. Is there a sponsor's remit?
2. Is there a business specification?
3. Has the project team published a description of the project in response to the above?
4. Is there a project definition document?
5. Does the project definition document include sections covering:
   - Definitions of terms
   - A synopsis of the sponsor's remit and business specification
   - Constraints on implementation (timescales, access to sites, limitations of consents, key milestones etc.)
   - Exclusions (related work which will not be done)
   - Applicable standards
   - Commercial conditions
   - Organisational structure (internal and external).
6. Are the principal points of contact between project and sponsor, and project and subcontractors/suppliers identified?
7. Is there a work breakdown structure (WBS)?
8. Are there scope statements in place for each package within the WBS?
9. Do the scope statements provide clear and unambiguous statements of the tasks to be performed?
10. Do the scope statements identify the applicable standards and constraints?
11. Do the scope statements allocate responsibility?
12. Do the scope statements allocate budgets for time and money?
13. Do the scope statements identify key points of contact both internal and external to the project?
14. Is there a network plan for the project?
15. Is every item of work in the scope statements identified on the network?
16. If there is more than one plan, do they interconnect?
17. Have the interface specifications been adhered to?
18. Where there are interconnects between packages implemented by different parties, are these interfaces defined by interface specifications which have been agreed by both sides?
19. Has the end date of the project been forecast by logic and task duration alone?
20. Is there a set of story boards which show the staging of the works at, say, quarterly intervals?
21. Have costs been prepared?
22. Are the costs based on the WBS? If not, what are they based on?
23. Has the basis for the costs been defined (assumptions, exclusions, inclusions etc.) and agreed?

24. Do the costs include contingencies for growth, change and claims?
25. Where costs have been provided by different parties, have they been made on the same basis (level playing field, double-accounting)?
26. Can the sponsor apply financial penalties?
27. Can the project apply financial penalties on subcontractors?
28. Is the project funded wholly or in part by non-client parties?
29. Is the non-client funding capped?
30. Does the client have to meet any cost increases?
31. Are there circumstances in which the client would have to pay money to the non-client parties?
32. Is the funding secured by contract?
33. Are the contractors working to firm prices?
34. Has a contractual risk allocation been made and agreed by all parties?
35. Does the project have staffing forecasts?
36. Do the staff exist?
37. Are there other, concurrent demands for the same staff from elsewhere?
38. Does the project have a central project office?
39. Do all the key managers work there?
40. Has a project manager been appointed?
41. Is there a design manager?
42. Is there a construction manager?
43. Is there a public consultation manager?
44. Is there a project accountant?
45. Does the project have cash flow forecasts?
46. Is there sufficient funding available?
47. Does the project have plant use forecasts?
48. Is the plant available?
49. Does the project have possession plans agreed with the operators?
50. Has an organisation structure been defined and communicated to all staff?
51. Does the project have regular and effective coordination meetings?
52. Have the scope statements been agreed by those who will do the work?
53. Have they been given start dates?
54. Do they have the necessary consents and routes to get to their place of work?
55. Have method statements been received from them?
56. Have the method statements been approved by the project?
57. Has time been allowed for mobilisation?
58. Has time been allowed for procurement?
59. Has time been allowed to plan and effect diversions?
60. Has the work started?
61. Have those doing the work provided more detailed programmes of the work?
62. Do the detailed programmes interface to the main project programme?
63. Do the detailed programmes compromise the project programme?
64. Have the contractors supplied method statements which are agreeable?
65. Are stages within a work package reviewed by experts before subsequent stages are started?
66. Is the work being done in accordance with written-down procedures?
67. Are regular written progress reports scheduled?

68. Does the project visit the works to check progress?
69. Does the project keep a project log?
70. Is there any way in which those working for the project can make withdrawals from the project accounts without approval?

List 6: The 12 I's and 1 R

This list is a favourite for identifying missing costs in a project estimate. These costs never appear on engineering drawings or site plans. Most are self-explanatory, but I have added some clarification where I think it may help.

	Possible cost	Comment
1	IT	
2	Income tax	
3	Interest	I know of one major project that was charged interest on negative cash flows
4	Insurances	This may pertain where the works of the project include something outside of ordinary business
5	Inflation	
6	Incentives	
7	In service cost	This is cost of spares and repair services. Some spares may not be always available and so a stock will have to be procured
8	In house services	High level overheads: board, legal, HQ
9	Inspection	Independent reviews. Testing services ... any service needed or imposed
10	Inconvenience	Compensations
11	Incomplete knowledge	Surveys and searches
12	Istorical	Costs from earlier studies or aborted works, the costs of which have not been written off but are awaiting a cost-collection number
13	Repair	Repairs that have to done to facilitate access to sites, safety assurance etc.

Appendix 2
The Example Risk Model

Summary Sheet

Project	Holsworth Railway Junction Project
risk model version	A 26th March 2011

The Estimate

general civils	£5,435,000
p way	£2,233,000
signalling	£2,800,000
traction power	£1,000,000
m&e	£430,000
telecomms	£375,000
-	-
project management cost	£1,600,000
IT services costs	£750,000
HQ costs	£1,250,000
spot cost	£15,873,000

The Plan

spot duration in weeks	121

From the Cost Risk Model

spread cost increment	#NAME?
total cost	#NAME?
cost after ownership analysis (say)	#NAME?

From the Time Risk Model

spread duration in weeks	#NAME?
start date	01-Mar-10
planned finish date	25-Jun-12
modelled finish date	#NAME?
spread/spot ratio	#NAME?
% increase in duration	#NAME?
completion of the design phase	#NAME?

Cost Risk Model

Example risk no	Primitive	Name	Synopsis	Modelling note	#1
1	Continuous Quantity	P way spread on Quantity	The quantity of rail track needed is a best guess. Between -1 and +1 km more could be required @ £1M per km but with probability reducing towards the limits.	*spread Quantity (continuous): quantity modelled as Triangular -1..0..1*	#NAME?
2	Discrete Quantity and Variable Price		Some existing trackside equipment cabinets may have to be moved to another site. Funding to move 6 has been included in the estimate but up to 4 more may have to be as well, depending on the final layout of the track. Each number (from 6 to 6+4) has an equal likelihood of occurrence.	spread Quantity (discrete):	#NAME?
		Cabinets spread on Quantity and Rate	The cost of relocating a cabinet depends on the number in the contract for the work, from a £40k per unit down for 6, reducing to £25k per unit for 10.	spread Rate (covarying):	#NAME?
3	Variable Productivity	Productivity Improvements	The project must pay the incumbent maintenance company to re-instate 20 lengths of overhead electrification cables on a time and materials basis. A working day costs £10,000 and either 1 or 2 or 3 lengths are fitted during it, with a relative frequency of occurrence of 20% to 50% to 30%. The outturn costs are thus variable but the estimate includes 20 days costs, and so there is an opportunity the outturn cost could be less.		cost per day

no. of lengths to fit

spot cost

		lengths	accumulator
production on day 1		#NAME?	#NAME?
production on day 2		#NAME?	#NAME?
production on day 3		#NAME?	#NAME?
production on day 4		#NAME?	#NAME?
production on day 5		#NAME?	#NAME?

#2	#3	Sample	Risk Data Output label	Selected risk data output value	Management action
		#NAME?	Risk 1: P way spread on Quantity	#NAME?	Surveys of possible alignments are scheduled to complete in Q3 and so by Q4 this risk should have reduced to an unavoidable error in measurement.
					However the final alignment may be a longer rather than a shorter and so the risk exposure should remain in the assessment in case it needs to be converted to a cost.
					No direct action by the RM is required.
#NAME?					The RM will - notify advise the estimator of the possible shortfall - notify the procurement manager of the need not to fix the number of cabinets in the PO at the moment - and request the lead engineer to investigate and resolve the uncertainty.
#NAME?		#NAME?	Risk 2: Cabinets spread on Quantity and Rate	#NAME?	
£10,000					The RM will advise the commercial manager of the opportunity and devise an incentive scheme that could be offered to the contractor. The decision whether or not to implement this will be for the Commercial Director to make.
20					
£200,000					

completion marker	day counter
-	1
#NAME?	2
#NAME?	3
#NAME?	4
#NAME?	5

Example risk no	Primitive	Name	Synopsis	Modelling note	#1
			production on day 6	#NAME?	#NAME?
			production on day 7	#NAME?	#NAME?
			production on day 8	#NAME?	#NAME?
			production on day 9	#NAME?	#NAME?
			production on day 10	#NAME?	#NAME?
			production on day 11	#NAME?	#NAME?
			production on day 12	#NAME?	#NAME?
			production on day 13	#NAME?	#NAME?
			production on day 14	#NAME?	#NAME?
			production on day 15	#NAME?	#NAME?
			production on day 16	#NAME?	#NAME?
			production on day 17	#NAME?	#NAME?
			production on day 18	#NAME?	#NAME?
			production on day 19	#NAME?	#NAME?
			production on day 20	#NAME?	#NAME?
4	Unresolved Option: One of Two	Track crossing for cables: unresolved option	The intention to install twelve under-track plastic cable ducts may be rejected as non-standard and instead 2 concrete culverts may have to be installed at a cost of £150k more.	unresolved option (permutation of 1 from 2), @ 50/50 for the cost difference.	#NAME?
5	Unresolved Option: M from N	Land take unresolved option	There is a probability that any of up to six pieces of land will need to be bought. Each piece of land has a different value. The Property Department say purchase will only be needed if a specific request to divert the railway away from a site of scientific interest is agreed. It is thought that there is 75% chance that the request will be denied.	unresolved option (permutation of N from 6) @ 25%	

Land Parcel No.	unique random key
#1	#NAME?
#2	#NAME?
#3	#NAME?
#4	#NAME?
#5	#NAME?
#6	#NAME?
rank	value of key
1	#NAME?
2	#NAME?
3	#NAME?
4	#NAME?

#2	#3	Sample	Risk Data Output label	Selected risk data output value	Management action
#NAME?	6				
#NAME?	7				
#NAME?	8				
#NAME?	9				
#NAME?	10				
#NAME?	11				
#NAME?	12				
#NAME?	13				
#NAME?	14				
#NAME?	15				
#NAME?	16				
#NAME?	17				
#NAME?	18				
#NAME?	19				
#NAME?	20				
		#N/A	Risk 3: Productivity Improvements	#NAME?	
		#NAME?	Risk 4: Track crossing for cables: unresolved option	#NAME?	
Cost of land parcel	Land Parcel No.				
#NAME?	#1				
#NAME?	#2				
#NAME?	#3				
#NAME?	#4				
#NAME?	#5				
#NAME?	#6				
value	Land Parcel No.				
#NAME?	#NAME?				
#NAME?	#NAME?				
#NAME?	#NAME?				
#NAME?	#NAME?				

Example risk no	Primitive	Name	Synopsis	Modelling note	#1
				5	#NAME?
				6	#NAME?
				index	accumulator
				0	£0
				1	#NAME?
				2	#NAME?
				3	#NAME?
				4	#NAME?
				5	#NAME?
				6	#NAME?
				permutation N	Cost
				#NAME?	#NAME?
6	Emerging Cost Chain: One Time Cost	Emerging cost chain for bridge replacement	The steel deck bridge at Lumley will need replacement (the cost of which is in the estimate) however the abutments are in a poor state of repair and might need replacement as well. The ground behind the one on the east side is water logged because the drains have failed and so draining and reinforcing may be needed.	emerging cost chain: 75% chance that new abutments will be needed @ £300k..£400k the pair, with a 75% chance that drain repairs and ground reinforcement costing £25k will needed on the east side.	3 cases: 1) replace abutments and do ground reinforcement: 75%*75% = 56.25% 2) replace abutments but no ground reinforcement: 75% * 25% = 18.75% 3) nothing required: 25%
7	Emerging Cost Chain: Multiple Cost, with time cross-link	Emerging cost chain for embankment slip	The embankment to the west has a history of slips: one in the last five years. Another slip may happen while the works are underway..	emerging cost chain: 20% chance of a slip during the duration of the project, incurring £250k…£500k of works and 4…6 week extension.	
				landslip occurs?	#NAME?
				no. of landslips	#NAME?
				constraint on no. of landslips	#NAME?
				no landslip	0
				first landslip	1
				second landslip	2
				third landslip	3
				fourth landslip	4

#2	#3	Sample	Risk Data Output label	Selected risk data output value	Management action
#NAME?	#NAME?				
#NAME?	#NAME?				
#NAME?		#NAME?	Risk 5: Land take unresolved option	#NAME?	
#NAME?		#NAME?	Risk 6: Emerging cost chain for bridge replacement	#NAME?	
£0	£0				
#NAME?	#NAME?				
#NAME?	#NAME?				
#NAME?	#NAME?				
#NAME?	#NAME?	#NAME?	Risk 7: Emerging cost chain for embankment slip	#NAME?	

Example risk no	Primitive	Name	Synopsis	Modelling note	#1
8	Random Failure	Compensation for Late Completions	20 overnight closures are planned for rural crossings for the commissioning of automatic barriers. Historically, work overruns in one in ten overnight closures, in which circumstances a £2500 fixed penalty has to be paid to the Local Authority for the inconvenience.	random failures	#NAME?
9	Price of Perfect Information	Public disbenefit	The public perceive the works are detrimental to area and will affect property values, quality of life etc.	too difficult to research. The Price of Perfect Information not worth paying.	
10	Strategic Risk	Project Cancelled	The scheme is incompatible with the intentions of Her Majesty's Opposition and it may be cancelled if there is a change of government.	Strategic risk. Not modelled.	
11	Safety Risk	Public Harm Risk	The works of the project must not alter the overall hazard to the public. Any works that do must be compensated by additional works that restore the status quo.	Safety risk.	A £50k sum has been added as a reserve to mitigate any perceived harm to the public.
12	Time dependent Cost: time risk model cross-link	PM charges are time dependent	Project Management charges are per accounting period for the duration of the project. If the project overruns, its cost will increase.	Time dependent cost. See Time Risk Model	

#2	#3	Sample	Risk Data Output label	Selected risk data output value	Management action
		#NAME?	Risk 8: Compensation for Late Completions	#NAME?	
		not modelled	Risk 9: Public disbenefit	not modelled	
		not modelled	Risk 10: Project Cancelled	not modelled	
		£50,000	Risk 11: Public Harm Risk	#NAME?	
		#NAME?	Risk 12: PM charges are time dependent	#NAME?	
	cost adjustment for risk	#NAME?			

Time Risk Model

	A	B	C	D	E	F	G	H	I/J	K	L	M	N	O
1	**SPOT VERSION**								**SPREAD i.e. RISK VERSION**					
2	Task No.	Task	start	predecessor task numbers	duration	finish	start date	finish date	Task No.	Task	start	predecessor task numbers	risk no.	primitive
3	1	prepare Business Case paper for authorisation for Feasibility and Outline Design	0	-		4 =C3+E3	=DATE (2010,3,1)	=$G$3+F3*7	1	prepare Business Case paper for authorisation for Feasibility and Outline Design	0	-		
4	2	obtain Business Case authorisation	=F3	1		1 =C4+E4	=$G$3+C4*7	=$G$3+F4*7	2	obtain Business Case authorisation	=V3	1		
5	3	appoint consultants	=F4	2		4 =C5+E5	=$G$3+C5*7	=$G$3+F5*7	3	appoint consultants	=V4	2	13	continuous spread duration
6	4	carry out Feasibility and Outline Design	=F5	3		24 =C6+E6	=$G$3+C6*7	=$G$3+F6*7	4	carry out Feasibility and Outline Design	=V5	3		
7	5	Sponsors assess Feasibility Study	=C6+4	4 SS+4w		8 =C7+E7	=$G$3+C7*7	=$G$3+F7*7	5	Sponsors assess Feasibility Study	=L6+4	4 SS+4w	14	discontinuous spread duration I
8	6	prepare Business Case paper for authorisation for Detailed Design and Implementation	=MAX(C6+10,F7)	4SS+10, 5		4 =C8+E8	=$G$3+C8*7	=$G$3+F8*7	6	prepare Business Case paper for authorisation for Detailed Design and Implementation	=MAX (L6+10,V7)	4SS+10, 5		
9	7	obtain Business Case authority	=F8	6		4 =C9+E9	=$G$3+C9*7	=$G$3+F9*7	7	obtain Business Case authority	=V8	6	22	multiple rework cycle
10	8	tender for services of consultants and contractors	=F9	7		16 =C10+E10	=$G$3+C10*7	=$G$3+F10*7	8	tender for services of consultants and contractors	=V9	7		
11	9	appoint consultants and contractors	=F10	8		8 =C11+E11	=$G$3+C11*7	=$G$3+F11*7	9	appoint consultants and contractors	=V10	8	21	part re-work cycle
12	10	carry out surveys and do detailed design: all disciplines	=F11	9		24 =C12+E12	=$G$3+C12*7	=$G$3+F12*7	10	carry out surveys and do detailed design: all disciplines	=V11	9		

P	Q	R #1	S #2	T #3	U duration	V finish	W start date	X finish date
synopsis	modelling note							
	spread duration				=RiskUniform(2,4)	=L3+U3	='summary sheet'!E19	=$G$3+V3*7
	spread duration				1	=L4+U4	=$G$3+L4*7	=$G$3+V4*7
The time required to appoint design consultants depends on how well the negotiations proceed.	spread duration: 4..12 weeks				=RiskUniform(4,12)	=L5+U5	=$G$3+L5*7	=$G$3+V5*7
					=RiskUniform(20,28)	=L6+U6	=$G$3+L6*7	=$G$3+V6*7
The duration allowed for Sponsors to review the Feasibility Study is 8 weeks. They might be able to review it in 4 weeks - an opportunity. Even better, because they have already reviewed earlier identical schemes, they might skip further review and authorise the project within a week.	spread duration: 4..8 weeks for a review, or one week if they proceed direct to authorisation. 75%/25% split.		=RiskDiscrete({1,2},{75,25})		=IF(R7'=1,RiskUniform(4,8),1)	=L7+U7	=$G$3+L7*7	=$G$3+V7*7
	spread duration				=RiskUniform(3,6)	=L8+U8	=$G$3+L8*7	=$G$3+V8*7
The application for funds to purchase new plant is likely to be rejected repeatedly by the Investment Committee, each rejection taking the form of a request for further information. The Committee sits every 4 weeks. It has been decided to plan for three re-applications on an assumption a fourth one will be successful. There is thought to be a 25% chance of success with the first application, 50% with a second, 15% with a third and 10% with a fourth.	multiple re-work cycle: 25% chance of an one cycle (4 weeks); 50% chance of two cycles; 15% of three and 10% of four.				=RiskDiscrete({4,8,12,16}, {25,50,15,10})	=L9+U9	=$G$3+L9*7	=$G$3+V9*7
	spread duration				=RiskUniform(12,16)	=L10+U10	=$G$3+L10*7	=$G$3+V10*7
A application for authorisation to let contracts may be rejected because a tightening of authorisation levels is anticipated. This will mean that the limits of delegated authority will be reduced and so one further higher level signature may be needed. This will not require an entire re-application though, and a signed memorandum confirming the approval will suffice.	part-rework cycle: 10% chance of a additional 25% of the planned time frame of the contract negotiation period.				=E11+(RiskDiscrete({1,0},{10,90}) *(E11*25%))	=L11+U11	=$G$3+L11*7	=$G$3+V11*7
	spread duration				=RiskUniform(24,30)	=L12+U12	=$G$3+L12*7	=$G$3+V12*7

13	11	clear work site of existing infrastructure	=F11 9	3	=C13+E13	=$G$3+C13*7	=$G$3+F13*7	11	clear work site of existing infrastructure	=V11 9	14	Compound Risk Duration
14	12	consult affected parties	=F6 4	24	=C14+E14	=$G$3+C14*7	=$G$3+F14*7	12	consult affected parties	=V6 4		
15	13	mobilise the piling contractor and construct piles	=C13+4 11SS+4	12	=C15+E15	=$G$3+C15*7	=$G$3+F15*7	13	mobilise the piling contractor and construct piles	=L13+RiskUniform(4,8) 11SS+4	17	Variable lag
16	14	mobilise the other contractors	=MAX(F12,F13, F14,F15) 10, 11, 12, 13	6	=C16+E16	=$G$3+C16*7	=$G$3+F16*7	14	mobilise the other contractors	=MAX (V12,V13, V14,V15) 10, 11, 12, 13		
17	15	erect signals	=F16	5	=C17+E17	=$G$3+C17*7	=$G$3+F17*7	15	erect signals	=V16	15	Spread duration based on historical data
18	16	do general civils	=F16 14	24	=C18+E18	=$G$3+C18*7	=$G$3+F18*7	16	do general civils	=V16 14	19	compound emerging task
19	17	-						17	-			
20	-	-						-	-			
21	18	do rail track civils	=C18+8 16SS+8	12	=C21+E21	=$G$3+C21*7	=$G$3+F21*7	18	do rail track civils	=L18+8 16SS+8		
22	19	do signalling installation mods	=F16 14	£12.00	=C22+E22	=$G$3+C22*7	=$G$3+F22*7	19	do signalling installation mods	=V16 14	23	interruption

Decommissioning a redundant signal box will take between 2 and 4 weeks to do. However the services of specialist technicians will be needed to terminate cables safely. They are involved in several other projects at the moment and there may be a 0-4 week delay obtaining their services, and even when they do become available, there is a chance access to the work site will be denied for a further 4 weeks while the necessary permissions are awaited.	Three durations are possible: Uniform 2 -4 weeks, or Uniform 0 - 4 weeks and then uniform 2 -4 weeks, or 4 weeks, then Uniform 0 -4 weeks and finally Uniform 2 -4 weeks. Each is modelled as having an equal probability of occurrence.	=RiskUniform (2,4)	=RiskUniform(0,4) +RiskUniform(2,4)	=4+RiskUniform(0,4) +RiskUniform(2,4)	=RiskOutput()+CHOOSE (RiskDiscrete({1,2,3},{1,1,1}), R13,S13,T13)	=L13+U13	=$G$3+L13*7	=$G$3+V13*7
						24 =L14+U14	=$G$3+L14*7	=$G$3+V14*7
Piling is scheduled to start 4 weeks after site clearance has started, however the rigs are being used on another project and their arrival on this project could be anytime between 4 and 8 weeks. The piling will then take 6..12 weeks to do depending on productivity rates.	fuzzy overlap: 4 to 8 weeks to get sub-contractors on site, then 8..12 weeks to do the work.				=RiskUniform(6,12)	=L15+U15	=$G$3+L15*7	=$G$3+V15*7
						6 =L16+U16	=$G$3+L16*7	=$G$3+V16*7
The project needs to erect ten signal posts. This is a common task and because it is often has to be done during short closures of the railway, contractors keep records of how long it takes to do the work. The range of times observed has a broad spread because the works were carried out at sites that have different access problems and because they were done in different weather conditions.	modelled as a compact accumulation of ten masts @ RiskDiscrete({1,2,3,4,5,6,7,8, 9,10},{0.02,0.03,0.04,0.18,0. 21,0.16,0.12,0.13,0.07,0.04}) days per mast. The relative probabilities are from the production records.				=(RiskDiscrete({1,2,3,4,5,6,7,8,9,10}, {0.02,0.03,0.04,0.18,0.21,0.16,0.12,0 .13,0.07,0.04}) +RiskDiscrete({1,2,3,4,5,6,7,8,9,10}, {0.02,0.03,0.04,0.18,0.21,0.16,0.12,0 .13,0.07,0.04}) +RiskDiscrete({1,2,3,4,5,6,7,8,9,10}, {0.02,0.03,0.04,0.18,0.21,0.16,0.12,0 .13,0.07,0.04}) +RiskDiscrete({1,2,3,4,5,6,7,8,9,10}, {0.02,0.03,0.04,0.18,0.21,0.16,0.12,0 .13,0.07,0.04}) +RiskDiscrete({1,2,3,4,5,6,7,8,9,10}, {0.02,0.03,0.04,0.18,0.21,0.16,0.12,0 .13,0.07,0.04}) +RiskDiscrete({1,2,3,4,5,6,7,8,9,10}, {0.02,0.03,0.04,0.18,0.21,0.16,0.12,0 .13,0.07,0.04}) +RiskDiscrete({1,2,3,4,5,6,7,8,9,10}, {0.02,0.03,0.04,0.18,0.21,0.16,0.12,0 .13,0.07,0.04}) +RiskDiscrete({1,2,3,4,5,6,7,8,9,10}, {0.02,0.03,0.04,0.18,0.21,0.16,0.12,0 .13,0.07,0.04}) +RiskDiscrete({1,2,3,4,5,6,7,8,9,10}, {0.02,0.03,0.04,0.18,0.21,0.16,0.12,0 .13,0.07,0.04}) +RiskDiscrete({1,2,3,4,5,6,7,8,9,10}, {0.02,0.03,0.04,0.18,0.21,0.16,0.12,0 .13,0.07,0.04}))/7	=L17+U17		
Public consultation may result in need to put up fencing….	emerging task: very likely so 75% chance of another 4..6 weeks work. The costs are in the estimate.	=RiskDiscrete({0,1},{25,75})		=RiskUniform(4,6)	=E18+(R18*IF($S18=1,S18,S1 9))+S20	=L18+U18	=$G$3+L18*7	=$G$3+V18*7
…or public consultation may result in need to construct noise bunds.	emerging task: very likely but as an alternative to fencing on 50/50 basis. Another 6..10 weeks. No extra costs incurred.	=RiskDiscrete({1,0},{50,50})		=RiskUniform(6,10)				
Interconnection to Risk 7 Emerging cost chain for embankment slip, from the cost risk model sheet..	emergent cost chain: 20% chance of a slip during the duration of the project, incurring 250k…500k of works and 4..6 week extension.	='cost risk model'!F68		=CHOOSE(R20+1,0,R iskUniform(4,6),RiskU niform(4,6)+RiskUnifor m(4,6),RiskUniform(4, 6)+RiskUniform(4,6)+R iskUniform(4,6),RiskU niform(4,6)+RiskUnifo rm(4,6)+RiskUniform(4, 6)+RiskUniform(4,6))				
	spread duration				=RiskUniform(10,20)	=L21+U21	=$G$3+L21*7	=$G$3+V21*7
Installation engineers booked for this task may have to be diverted to a major systems commissioning task on project X that is also planned to happen at the same time.	interruption: low probability. but project X will take priority on resources. If it does then this commissioning will have to be deferred two weeks.				=E22+RiskDiscrete({1,0},{25,75})*2	=L22+U22	=$G$3+L22*7	=$G$3+V22*7

23	20	commission systems	=MAX 15, 16, 18, 19 (F17,F18,F21,F22)	4	=C23+E23	=$G$3+C23*7	=$G$3+F23*7		20	commission systems	=MAX 15, 16, (V17,V18, 18, 19 V21,V22)	20	re-work cycle
24	21	re-instate overhead wiring	=F23 20	2	=C24+E24	=$G$3+C24*7	=$G$3+F24*7		21	re-instate overhead wiring	=V23 20	26	variable productivity
25	.	.										
26	.	.										
27	.	.										
28	.	.										
29	.	.										
30	.	.										
31	.	.										
32	.	.										
33	.	.										
34	.	.										
35	.	.										
36	.	.										
37	.	.										
38	.	.										
39	.	.										
40	.	.										
41	.	.										
42	.	.										
43	.	.										
44	.	.										
45	.	.										
46	22	evaluation period	=F24 21	4	=C46+E46	=$G$3+C46*7	=$G$3+F46*7		22	evaluation period	=V24 21	24	multiple task interruption
47	23	pack up site	=F46 21	4	=C47+E47	=$G$3+C47*7	=$G$3+F47*7		23	pack up site	=V46 22	25	disruption
48	24	landscape the site	=F47 23	0	=C48+E48	=$G$3+C48*7	=$G$3+F48*7		24	landscape the site	=V47 23	18	simple emerging task
49	25								25			
50	26								26			
51					spot >	=F47						
52												

The chosen type of non-interruptible backup generators may fail their full load tests. If they do they will have to be up-rated by re-instating features that were removed by in a Value Engineering exercise. The commissioning task will have to be repeated if this happens. In the opinion of the electrical engineers there is a 50% chance of this happening.	re-work cycle: 50% chance that the commissioning will have to be repeated.					=E23+(RiskDiscrete({0,1},{50,50}))*E23	=L23+U23 =$G$3+L23*7 =$G$3+V23*7	
Re-instatement of overhead wiring sections is planned take place over 10 days: 20 sections at a rate of 2 sections per day. The historical record shows that early days are susceptible to plant breakdowns and logistic mishandlings but that productivity does improve to a sustained average of 4 sections per day.	productivity growth: each day is modelled with production rate of 0 to 6 sections but the probability of each outcome biased towards 0 for the first working week, thereafter altering steadily towards 6 sections.					=VLOOKUP(1,T26:U45,2,FALSE)/7	=L24+U24 =$G$3+L24*7 =$G$3+V24*7	

day no. label	production	acc. Production	quota achieved?	day no.
day 1	=RiskDiscrete({0,1,2,3,4,5,6},{40,30,20,10,0,0,0})	=R26	=IF(S26<20,0,1)	1
day 2	=RiskDiscrete({0,1,2,3,4,5,6},{40,30,20,10,0,0,0})	=S26+R27	=IF(S27<20,0,1)	2
day 3	=RiskDiscrete({0,1,2,3,4,5,6},{30,40,20,10,0,0,0})	=S27+R28	=IF(S28<20,0,1)	3
day 4	=RiskDiscrete({0,1,2,3,4,5,6},{20,40,30,10,0,0,0})	=S28+R29	=IF(S29<20,0,1)	4
day 5	=RiskDiscrete({0,1,2,3,4,5,6},{20,30,40,10,0,0,0})	=S29+R30	=IF(S30<20,0,1)	5
day 6	=RiskDiscrete({0,1,2,3,4,5,6},{10,30,40,20,0,0,0})	=S30+R31	=IF(S31<20,0,1)	6
day 7	=RiskDiscrete({0,1,2,3,4,5,6},{10,20,40,20,10,0,0})	=S31+R32	=IF(S32<20,0,1)	7
day 8	=RiskDiscrete({0,1,2,3,4,5,6},{10,10,30,30,10,10,0})	=S32+R33	=IF(S33<20,0,1)	8
day 9	=RiskDiscrete({0,1,2,3,4,5,6},{10,10,30,20,10,10,10})	=S33+R34	=IF(S34<20,0,1)	9
day 10	=RiskDiscrete({0,1,2,3,4,5,6},{10,10,30,20,10,10,10})	=S34+R35	=IF(S35<20,0,1)	10
day 11	=RiskDiscrete({0,1,2,3,4,5,6},{10,10,30,20,10,10,10})	=S35+R36	=IF(S36<20,0,1)	11
day 12	=RiskDiscrete({0,1,2,3,4,5,6},{10,10,30,20,10,10,10})	=S36+R37	=IF(S37<20,0,1)	12
day 13	=RiskDiscrete({0,1,2,3,4,5,6},{10,10,30,20,10,10,10})	=S37+R38	=IF(S38<20,0,1)	13
day 14	=RiskDiscrete({0,1,2,3,4,5,6},{10,10,30,20,10,10,10})	=S38+R39	=IF(S39<20,0,1)	14
day 15	=RiskDiscrete({0,1,2,3,4,5,6},{10,10,30,20,10,10,10})	=S39+R40	=IF(S40<20,0,1)	15
day 16	=RiskDiscrete({0,1,2,3,4,5,6},{10,10,30,20,10,10,10})	=S40+R41	=IF(S41<20,0,1)	16
day 17	=RiskDiscrete({0,1,2,3,4,5,6},{10,10,30,20,10,10,10})	=S41+R42	=IF(S42<20,0,1)	17
day 18	=RiskDiscrete({0,1,2,3,4,5,6},{10,10,30,20,10,10,10})	=S42+R43	=IF(S43<20,0,1)	18
day 19	=RiskDiscrete({0,1,2,3,4,5,6},{10,10,30,20,10,10,10})	=S43+R44	=IF(S44<20,0,1)	19
day 20	=RiskDiscrete({0,1,2,3,4,5,6},{10,10,30,20,10,10,10})	=S44+R45	=IF(S45<20,0,1)	20

The evaluation period requires 4 weeks of ambient temperatures >0°C. If the task happens to fall in January or February there is a 10% chance that any working day will experience a temperature lower than this, in which case an additional day will need to be added to the overall duration of the task.	modelled as a 10% chance of any working day being lost during January and February in 5-day calendar.	=VLOOKUP(W46,'5 day calendar'!$A$2:$C$1097,6,FALSE)	=R46+(E46*7)	=VLOOKUP(S46,'5 day calendar'!$F$2:$G$1097,2,FALSE)	=(X46-W46)/7	=L46+U46	=INT($G$3+L46*7)	=T46
The clients may wish to move on to the site two weeks after the evaluation period, in case the last half of the time required to clear the site will be disrupted and take up to 50% longer.	disruption after 50% completion, with up to 100% extension of the remaining balance of time.				=(E47*50%)+(E47*(1-50%)*(1+RiskUniform(0%,100%)))	=L47+U47 =$G$3+L47*7 =$G$3+V47*7		
There is a possibility that some landscaping work will have to be done to make the site acceptably attractive.	emerging task. Very likely to be a 75% chance of another 1..4 weeks.	=RiskDiscrete({1,0},{75,25})	=RiskUniform(1,4)		=R48*S48	=L48+U48 =$G$3+L48*7 =$G$3+V48*7		

spread > =V47
% increase > =V51/F51-1

5 Day Calendar

Date	Day	Test for working days in the week	Test for January or February	Risk: a working day in either January or February on which the ambient temperature falls below 0°C becomes a non-working day	Total working days after risk is modelled	Date (repeat)
01-Jan-12	Sun	0	1	0	0	01-Jan-12
02-Jan-12	Mon	1	1	#NAME?	#NAME?	02-Jan-12
03-Jan-12	Tue	1	1	#NAME?	#NAME?	03-Jan-12
04-Jan-12	Wed	1	1	#NAME?	#NAME?	04-Jan-12
05-Jan-12	Thu	1	1	#NAME?	#NAME?	05-Jan-12
06-Jan-12	Fri	1	1	#NAME?	#NAME?	06-Jan-12
07-Jan-12	Sat	0	1	0	#NAME?	07-Jan-12
08-Jan-12	Sun	0	1	0	#NAME?	08-Jan-12
09-Jan-12	Mon	1	1	#NAME?	#NAME?	09-Jan-12
10-Jan-12	Tue	1	1	#NAME?	#NAME?	10-Jan-12
11-Jan-12	Wed	1	1	#NAME?	#NAME?	11-Jan-12
12-Jan-12	Thu	1	1	#NAME?	#NAME?	12-Jan-12
13-Jan-12	Fri	1	1	#NAME?	#NAME?	13-Jan-12
14-Jan-12	Sat	0	1	0	#NAME?	14-Jan-12
15-Jan-12	Sun	0	1	0	#NAME?	15-Jan-12
16-Jan-12	Mon	1	1	#NAME?	#NAME?	16-Jan-12
17 Jan-12	Tue	1	1	#NAME?	#NAME?	17-Jan-12
18-Jan-12	Wed	1	1	#NAME?	#NAME?	18-Jan-12
19-Jan-12	Thu	1	1	#NAME?	#NAME?	19-Jan-12
20-Jan-12	Fri	1	1	#NAME?	#NAME?	20-Jan-12
21-Jan-12	Sat	0	1	0	#NAME?	21-Jan-12
22-Jan-12	Sun	0	1	0	#NAME?	22-Jan-12
23-Jan-12	Mon	1	1	#NAME?	#NAME?	23-Jan-12
24-Jan-12	Tue	1	1	#NAME?	#NAME?	24-Jan-12
25-Jan-12	Wed	1	1	#NAME?	#NAME?	25-Jan-12
26-Jan-12	Thu	1	1	#NAME?	#NAME?	26-Jan-12
27-Jan-12	Fri	1	1	#NAME?	#NAME?	27-Jan-12
28-Jan-12	Sat	0	1	0	#NAME?	28-Jan-12
29-Jan-12	Sun	0	1	0	#NAME?	29-Jan-12
30-Jan-12	Mon	1	1	#NAME?	#NAME?	30-Jan-12
31-Jan-12	Tue	1	1	#NAME?	#NAME?	31-Jan-12
01-Feb-12	Wed	1	1	#NAME?	#NAME?	01-Feb-12
02-Feb-12	Thu	1	1	#NAME?	#NAME?	02-Feb-12
03-Feb-12	Fri	1	1	#NAME?	#NAME?	03-Feb-12
04-Feb-12	Sat	0	1	0	#NAME?	04-Feb-12
05-Feb-12	Sun	0	1	0	#NAME?	05-Feb-12
06-Feb-12	Mon	1	1	#NAME?	#NAME?	06-Feb-12
07-Feb-12	Tue	1	1	#NAME?	#NAME?	07-Feb-12
08-Feb-12	Wed	1	1	#NAME?	#NAME?	08-Feb-12
09-Feb-12	Thu	1	1	#NAME?	#NAME?	09-Feb-12
10-Feb-12	Fri	1	1	#NAME?	#NAME?	10-Feb-12

Date	Day	Test for working days in the week	Test for January or February	Risk: a working day in either January or February on which the ambient temperature falls below 0°C becomes a non-working day	Total working days after risk is modelled	Date (repeat)
11-Feb-12	Sat	0	1	0	#NAME?	11-Feb-12
12-Feb-12	Sun	0	1	0	#NAME?	12-Feb-12
13-Feb-12	Mon	1	1	#NAME?	#NAME?	13-Feb-12
14-Feb-12	Tue	1	1	#NAME?	#NAME?	14-Feb-12
15-Feb-12	Wed	1	1	#NAME?	#NAME?	15-Feb-12
16-Feb-12	Thu	1	1	#NAME?	#NAME?	16-Feb-12
17-Feb-12	Fri	1	1	#NAME?	#NAME?	17-Feb-12
18-Feb-12	Sat	0	1	0	#NAME?	18-Feb-12
19-Feb-12	Sun	0	1	0	#NAME?	19-Feb-12
20-Feb-12	Mon	1	1	#NAME?	#NAME?	20-Feb-12
21-Feb-12	Tue	1	1	#NAME?	#NAME?	21-Feb-12
22-Feb-12	Wed	1	1	#NAME?	#NAME?	22-Feb-12
23-Feb-12	Thu	1	1	#NAME?	#NAME?	23-Feb-12
24-Feb-12	Fri	1	1	#NAME?	#NAME?	24-Feb-12
25-Feb-12	Sat	0	1	0	#NAME?	25-Feb-12
26-Feb-12	Sun	0	1	0	#NAME?	26-Feb-12
27-Feb-12	Mon	1	1	#NAME?	#NAME?	27-Feb-12
28-Feb-12	Tue	1	1	#NAME?	#NAME?	28-Feb-12
29-Feb-12	Wed	1	1	#NAME?	#NAME?	29-Feb-12
01-Mar-12	Thu	1	0	1	#NAME?	01-Mar-12
02-Mar-12	Fri	1	0	1	#NAME?	02-Mar-12
03-Mar-12	Sat	0	0	0	#NAME?	03-Mar-12
04-Mar-12	Sun	0	0	0	#NAME?	04-Mar-12
05-Mar-12	Mon	1	0	1	#NAME?	05-Mar-12
06-Mar-12	Tue	1	0	1	#NAME?	06-Mar-12
07-Mar-12	Wed	1	0	1	#NAME?	07-Mar-12
08-Mar-12	Thu	1	0	1	#NAME?	08-Mar-12
09-Mar-12	Fri	1	0	1	#NAME?	09-Mar-12
10-Mar-12	Sat	0	0	0	#NAME?	10-Mar-12
11-Mar-12	Sun	0	0	0	#NAME?	11-Mar-12
12-Mar-12	Mon	1	0	1	#NAME?	12-Mar-12
13-Mar-12	Tue	1	0	1	#NAME?	13-Mar-12
14-Mar-12	Wed	1	0	1	#NAME?	14-Mar-12
15-Mar-12	Thu	1	0	1	#NAME?	15-Mar-12
16-Mar-12	Fri	1	0	1	#NAME?	16-Mar-12

Error Function Model

on-site contractor's task no.	planned duration in lapsed weeks	actual duration in lapsed weeks	normalised weight
PH004-1	9.6	9.7	1.0138
PH004-2	1.3	1.2	0.9282
PH004-3	7.0	7.4	1.0568
PH004-4	8.0	8.2	1.0181
PH004-5	9.6	10.5	1.0916
PH004-6	16.0	19.2	1.1988
PH004-7	12.8	12.7	0.9955
PH004-8	12.9	11.8	0.9184
PH004-9	11.8	13.8	1.1697
PH004-10	18.9	19.2	1.0203
PH004-11	5.6	7.0	1.2380
PH004-12	13.9	13.0	0.9390
PH004-13	10.7	10.6	0.9898
PH004-14	3.2	3.3	1.0302
PH004-15	10.7	11.6	1.0855
PH004-16	1.9	2.3	1.1799
PH004-17	4.1	4.1	1.0187
PH004-18	12.5	12.1	0.9724
PH004-19	14.3	16.3	1.1374
PH004-20	14.9	16.1	1.0783
PH004-21	1.8	1.9	1.0285
PH004-22	3.3	3.2	0.9808
PH004-23	6.3	6.1	0.9638
PH004-24	3.0	3.5	1.1553
PH004-25	2.5	2.8	1.1367
PH004-26	11.1	11.7	1.0508
PH004-27	2.4	2.5	1.0564
PH004-28	4.2	4.4	1.0632
PH004-29	17.7	19.5	1.1019
PH004-30	8.9	8.8	0.9855
PH004-31	16.3	17.6	1.0783
PH004-32	12.5	14.7	1.1716
PH004-33	12.3	12.7	1.0327
PH004-34	16.1	16.8	1.0448
PH004-35	19.0	20.1	1.0611
PH004-36	18.1	20.4	1.1249
PH004-37	15.8	15.5	0.9805
PH004-38	16.5	18.7	1.1308

PH004-39	18.0	17.7	0.9831
PH004-40	17.6	22.0	1.2503
PH004-41	7.2	7.5	1.0414
PH004-42	19.6	19.9	1.0161
PH004-43	16.4	17.1	1.0452
PH004-44	6.2	6.3	1.0099
PH004-45	15.2	15.0	0.9891
PH004-46	7.4	7.4	0.9986
PH004-47	16.8	16.7	0.9909
PH004-48	14.1	14.2	1.0102
PH004-49	12.5	15.4	1.2296
PH004-50	7.6	8.1	1.0600
PH004-51	4.9	6.0	1.2114
PH004-52	19.1	18.4	0.9628
PH004-53	1.9	1.9	1.0145
PH004-54	1.1	1.2	1.0223
PH004-55	12.4	13.4	1.0849
PH004-56	5.6	5.4	0.9494
PH004-57	10.7	11.4	1.0613
PH004-58	2.3	2.7	1.1396
PH004-59	17.4	18.8	1.0782
PH004-60	19.6	21.5	1.0945
PH004-61	17.6	19.0	1.0767
PH004-62	3.3	3.3	0.9929
PH004-63	8.8	8.3	0.9390
PH004-64	13.1		
PH004-65	13.0		
PH004-66	13.4		
PH004-67	19.9		
PH004-68	8.9		
PH004-69	7.8		
PH004-70	16.7		
PH004-71	8.2		
PH004-72	14.2		
PH004-73	2.0		
PH004-74	5.8		
PH004-75	1.7		
PH004-76	19.5		
PH004-77	3.5		
PH004-78	15.3		
PH004-79	14.2		
PH004-80	16.1		

PH004-81	9.2
PH004-82	9.5
PH004-83	18.5
PH004-84	19.9
PH004-85	3.3
PH004-86	11.5
PH004-87	16.2
PH004-88	5.8
PH004-89	18.6
PH004-90	5.6
PH004-91	3.5
PH004-92	8.9
PH004-93	16.7
PH004-94	3.4
PH004-95	9.0
PH004-96	5.1
PH004-97	12.7
PH004-98	19.6
PH004-99	8.5
PH004-100	5.7

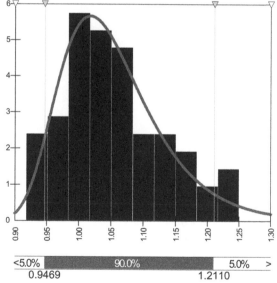

ExtValue(1.018113, 0.064932)

Grimbly Railway Junction Re-modelling (Spreadsheet)

The author has made a complete Excel spreadsheet available to support this book: *Grimbly Railway Junction Re-modelling*. This includes spreadsheet models (as separate tabs) to cover the cost risk model, the time risk model, five-day calendar and the error function model.

Readers may request a free copy of this spreadsheet to be supplied via e-mail by visiting www.gowerpublishing.com/grimblyspreadsheet

Index

Printed and bound by CPI Group (UK) Ltd, Croydon, CR0 4YY

18/10/2024

01776204-0007